INDIA'S SPECIAL FORCES

History and Future of Indian Special Forces

INDIA'S SPECIAL FORCES
History and Future of Indian Special Forces

by

Lt Gen P C Katoch, PVSM, UYSM, AVSM, SC (Retd)
&
Saikat Datta

(Established 1870)

United Service Institution of India, New Delhi

Vij Books India Pvt Ltd
New Delhi (India)

Published by

Vij Books India Pvt Ltd
(Publishers, Distributors & Importers)
2/19, Ansari Road, Darya Ganj
New Delhi - 110002
Phones: 91-11-43596460, 91-11- 47340674
Fax: 91-11-47340674
e-mail : vijbooks@rediffmail.com
web: www.vijbooks.com

Copyright © 2013, United Service Institution of India, New Delhi

ISBN: 978-93-82573-97-5

"Nothing is Permanent", says Bhagwad Gita.

implying

Only Change is Constant

India's Special Forces and Concept of their Employment
must change with changing geostrategic realities.

Dedication

This book is dedicated to all the men who ever associated or served in the Indian Special Forces and their families.

Authors Note on Sources

Writing about Special Forces is a delicate subject since very few records are publicly available. The bulk of the reporting for this book has come from interactions with people who were associated with the Special Forces. Where certain books or papers have been used, they have been cited accordingly.

Contents

Section III – The Future

Foreword

The nature of war has undergone dramatic changes in the past decade and countries are striving to ensure that their military is combat ready at all times to fight the right kind of war in respect of their concepts, organization, equipment, and training. We are also similarly engaged in the context of our security environment. Many of the wars today are being fought by proxy and the methods employed by both sides are un-conventional. A lesson from the recent Afghan and Iraq Wars is that 'victory is a chimera; counting on today's enemy to yield in the face of superior force makes about as much sense as buying lottery tickets to pay the mortgage - you have to be really lucky'. In most of the conflicts today pure military solutions simply do not exist and neither can the military kill their way out.

In the new era of conflict Special Forces have a major role to play, they give their governments a variety of options like controlling the escalation, overt or covert operations with deniability, reaching the adversary's vulnerable areas and limiting collateral damage when targeting critical strategic points. Special Forces has been an idea whose time had long come, however it is only in recent past that our military leaders and higher defence authorities have recognized their vast potential. It is not as easy as introducing another weapon in our armoury, it has to be regarded as a Weapon of War and there has been very little written about the concepts, organization and employment of our Special Forces to fit that role. This book titled Indian Special Forces has come at the right time and there could not be a more appropriate person to author than Lt Gen Prakash Katoch. His first-hand experience of over two years as Commanding Officer of 1 Special Forces Battalion in actual operations and his subsequent experience at higher echelons of Special Forces management make him the ideal person to write this treatise.

I had the privilege of having all the three original Special Forces battalions, 1 Para Commando, 9 Para Commando and 10 Para Commando as part of the Indian Peace Keeping Force (IPKF) under my command. They were then designated as Para Commandos and we had no doctrine on the employment of Special Forces; we had to learn it 'on the job'. Indeed initially most of the division and brigade commanders tended to regard them as an elite infantry additive to their battalions and tended to use them accordingly as regular infantry for the more difficult and dangerous tasks. It was difficult for many commanders to differentiate the role of Special Forces from that of para commandos and we had to put a stop to that straight away, and ultimately over two years of continuous operations the concept of their employment was finely honed and it had a defining impact on the Sri Lanka Campaign. One aspect I learnt in the campaign above all else was that the Special Forces are a strategic weapon to be used at the right time to tilt the battle in your favor.

This book is a holistic study of the Special Forces tracing their origin from the para-commandos to their ultimate reincarnation as Special Forces. The chapter on their future role and employment is highly insightful. I recommend the book for all students of conflict studies and analysts who have an interest in India's security issues.

-Lt Gen AS Kalkat, SYSM, PVSM, AVSM, VSM (Retd),
Former Overall Force Commander and GOC IPKF,
and later Army Commander ARTRAC

Preface

Special Forces are usually shrouded in half mysteries, their aura and lure overpowering. India is no different. In the late Eighties you could see security guards posted outside shops in Connaught Place, New Delhi with rakish maroon berets sporting 'Special Forces' or 'Commando' shoulder titles. While some tasks of Special Forces, commando forces and airborne forces may overlap at times that axiomatically is acknowledged globally, there has been some confusion in India in distinguishing between these forces; mostly in media but at times even in some military circles. Two decades back there was news of a Special Forces Regiment having been formed in the Indian Army and only a couple of months later disbandment of this fledgling Regiment. The Navy and Air Force established their own Special Forces in time, as also some other organizations came up tasked with special missions. That apart, periodically, mention of Special Forces keeps coming up in the media, mostly during conflict situations, successes in counter insurgency and anti-infiltration operations or instances like the spectacular US Special Forces raid that killed Osama-bin-Laden deep inside Pakistani territory, with questions aired as to capabilities of our own Special Forces to undertake similar missions. In recent past, media has reported one of the important recommendations of the Prime Minister appointed Naresh Chandra Committee for establishment of a separate Special Forces Command. A very careful analysis is needed as to why India needs Special Forces, in what measure and what form especially in an environment of strategic ambiguity in the absence of a national security strategy, national security objectives undefined and blending of 'military' with 'diplomacy' yet to be achieved.

Acknowledgements

First of all, I would like to thank Lieutenant General PK Singh, PVSM, AVSM (Retd), Director, United Service Institution (USI) of India and my erstwhile Army Commander under whom I had the privilege to command a Strike Corp, for being the very inspiration behind this book and in awarding me the Field Marshal KM Cariappa Chair of Excellence for the year 2011-2012. He was also kind in sponsoring me for the Middle East Special Operations Commanders Conference (MESOC 2012) held at Amman, Jordan in May 2012, which provided me the opportunity to interact with numerous special operations commanders and operatives and know about Special Forces operations in Iraq, Afghanistan and rest of the Middle East. I would also like to place on record the guidance provided by Major General YK Gera (Retd), Consultant (Research), USI in undertaking this research.

If we were to list out names of each and every individual that has contributed to this book, it would actually take a few pages. There may also be a problem, as in all militaries, if blunt views of individuals who are still in uniform are brought out in print attributing their names to the statements. Extensively travelling to numerous places in the past one year, we (Mr Saikat Datta and I as co-authors of this book) could interview and discuss all issues related to Special Forces with scores of individuals (sometimes even in groups), both serving and veterans of Special Forces, plus outstanding personalities (including Army Commanders and General officers) who had not personally served in Special Forces but had commanded these forces in conflict situations, counter insurgency, no war no peace and had contributed to development of Special Forces. Those that could not be met personally were contacted telephonically and / or through e-mail and their inputs taken including personal anecdotes and combat experiences, which have been included in the book in one form or another. The interaction included all ranks of some Special Forces units – officers, junior commissioned officers and other ranks. We also had the opportunity to interact with battle hardened veterans of 1965 and 1971 wars, operations under IPKF in Sri Lanka, Kargil conflict, counter insurgency operations and experiences in Siachen Glacier area. Over the past several years both of us having written prolifically on Special Forces and participated in seminars and given talks on the issue both at national and international levels. We have endeavored to incorporate those experiences in this book as well. We would, therefore, like to thank all those who spared their time and gave us their invaluable views that helped us compile this book.

I would like to place on record my special thanks to Mr Saikat Datta (co-author of this book) whom I have known for many years especially because of his intense and vast interest, and research undertaken in the subject of Special Forces. In 2011, I also had the privilege of running a special operations capsule for Force 1 in Mumbai along with him.

And finally, it is my wife Kamini, son Anuj and daughter-in-law Deepika who gave me their whole hearted support in this venture and my grateful thanks to them.

-Prakash Katoch

This book wouldn't have been possible but for one man who spent the better part of his life trying to improve our Special Forces. So my first acknowledgement among many others, must go to Lt Gen P C Katoch, a man whom I have come to respect immensely and look up to. Without him the Indian Special Forces would have probably been in a much worse shape than what they are in. With 40 years under his belt as a Special Forces officer, Lt Gen Katoch validated his concepts in a variety of operations, from counter-terrorism to counter-insurgency and conventional battle. He was a team commander, a commanding officer and later on a formation commander who spent the best of years of his life in service of his country and the Special Forces. The rest were spent in battling bureaucratic battles in South Block and Army Headquarters trying to make an unwilling military and bureaucratic hierarchy understand the importance of Special Forces. Besides that, his immense patience and guidance has been of tremendous value to me.

I also owe this book to a young boy who is perhaps the embodiment of our future. This is my little son Angad, who always reminds me how wonderful our future will be. Thanks are also due to my wife Priyamvada for her support, my friend Anuradha Raman for her encouragement and belief. The two of the best editors I worked under also deserve a great deal of thanks. The freedom they gave me at different stages of my career helped me to study the evolution of Special Forces and also write on them between crazy newsroom deadlines. So thank you Mr Vinod Mehta and Mr Aditya Sinha for shaping me in so many ways; I shall forever be in your debt.

This book wouldn't have been possible without the United Service Institution of India, which has shaped generations of some of the finest minds in our military.

Finally, the biggest thanks go to all the men who served or are serving in the Indian Special Forces. They are the prime reason why I worked on this book for so long. It is their contribution and sacrifice that has shaped this book and I owe them a debt of gratitude as an author and as a citizen.

Some people I can name – Maj Abhay Sapru (Retd), Col Jaideep Sengupta (Retd), Lt Gen Hardev Singh Lidder (Retd), Lidder, Maj Gen Dalvir Singh (Retd), Lt Gen Rustom Nanavatty (Retd), Air Cmde Jasjit Singh (Retd), Lt Gen Vijay Oberoi (Retd), Lt Gen Tej Pathak (Retd), Maj Anil Malik (Retd), Maj Dalip Bhalla (Retd), Col Alok Jha (Retd), the family of Maj Uday Singh who was killed in action, Col N S Rathore (Retd), Brig Ashok Taskar (Retd), Col Roshan Lal (Retd), Col V B Shinde (Retd), Major General R P Limaye (Retd) and Mr Sadanand Date, IPS are just some of the many, many people who came forward willingly to share their stories and personal testimonies. Besides them are hundreds of others who chose to step away from the limelight and share their stories, views and testimonies anonymously just because they loved the Special Forces more than I ever could. This book is my personal tribute to all of you. May you all live long and prosper.

-Saikat Datta

Introduction

To date, there is no book that traces the travails of the Special Forces of India albeit books have been authored earlier aimed at equating Airborne Forces with Special Forces and in one instance covering the history and exploits of one particular Special Forces unit. This book is attempted as a critique of Special Forces as they have emerged in India, as related to counterparts in all important armed forces around the world, emphasizing the concept as it has developed in our Army and the current name it has come to acquire. It delves into the creation of Special Forces in India; evolution, development through the years and why they are called Special Forces. What separates them from what was earlier being referred to as Commando Forces, Commando Operations or Special Missions? Were the operations or special missions being conducted by our Commando Battalions actually beyond the capabilities of the Infantry Battalions? Did the special missions by Commando Battalions differ in surgical operations, in their swiftness, the scale of operations which necessarily was small, their reach longer and an equally speedy exit? Did the commando elements continue to dominate their limited sphere at the tactical levels of units and formations?

Faced with cross border terrorism since decades, a host of insurgencies and our enemies' hell bent to wage asymmetric wars on us including through irregular forces employed within our own territory, have we made any effort at all to create necessary deterrence, if at all? Have we not been at serious disadvantage vis-à-vis China and Pakistan with the threat magnifying from a single front at one time to two-front, and now two-and-a-half front? Employment till now of our nascent Special Forces have been analyzed including whether our Special Forces have actually been employed or used as Special Forces or primarily used in counter insurgency operations for which we have any number of other units available. The book brings out whether a rare resource like Special Forces should or should not be employed for such missions that can be performed by a host of other groups organized under various entities of the government barring exceptions made if there is a hard nut to crack and which requires special techniques and tactics that are beyond the capability of regular units. The internet tells you India has a plethora of Special Forces including within the three Services that have been created to cater for specific tasks pertaining to their organizations and within their limited roles. They are also controlled by their own heads of organizations and departments; some under the Ministry of Defence (MoD), Ministry of Home Affairs (MHA) or Cabinet Secretariat. Can they really be brought together or their elements used as and when required by any of these constituents and if so, to what effect especially with their levels of training, equipment, tactics, and command and control all differing. Will the various departments let go of their resources easily, given our egos and turf battles or will the vested interests continue to stall and deter such effort. Will combining resources and inter-operability be feasible at all and if so, can it extend beyond the three Services? What magnitude of Special Forces does India require? With our tendency to numerically go on expanding our Special Forces numbering those of the USA, where are we in terms of resources and technology compared to them? Would it be more prudent to have smaller Special Forces with effective reach within areas of 'our' strategic interest? Should we not optimize our Special Forces potential with available technology and indigenous content? What deliberations are required in selection of manpower, equipping, training and most importantly command and control? In the case of Army Special Forces, there has been a tendency to divide the entire Special Forces effort by placing them under different Commands, whereby they virtually became Command troops consequently affecting any planning and deployment at strategic level. In the backdrop of 21st Century threats, what should be the Special Forces structure in India, their concept of employment and doctrine? These are the questions this book has attempted to answer. Wish you a happy reading !

Section – I
The Para Commando Years
(1965-1990)

1

Meghdoot Force

By November 1962, Prime Minister Pandit Jawaharlal Nehru was a broken man, still recovering from the bitter defeat at the hands of the Chinese a month ago. His 'forward policy' had failed, his Defence Minister had resigned and his favourite General, B M Kaul was consigned to ignominy. The Intelligence Bureau (IB) had failed to gauge Chinese intentions leading to the shock of 1962 and its Director, B M Mullick was busy trying to stave off demands to create a separate agency dedicated for gathering external intelligence.

While there aren't any public records of this period available, the CIA reportedly contacted Mullick and offered help that would help shore up Indian defences against "communist China". The CIA was falling back on a similar experiment they had successfully conducted in South Asia over a decade ago, when they helped the Pakistanis in an effort to shore up their defences against the Soviets. In 1954, the CIA had contacted the General Headquarters in Rawalpindi and convinced Field Marshal Ayub Khan that a commando unit, capable of "behind-enemy-lines activities" would help them if the Soviets were to cross the Hindu Kush and attack Pakistan. For Ayub, this was an opportunity to create a force that would be sanctioned against a possible Soviet invasion, but would be more suitable for any future wars against India.

For the CIA in 1962, then working under President John F Kennedy, this was a valuable opportunity to deal with India, which had been consistently pushing the non-aligned alternative for decades. Trainers from American Special Forces helped create the nucleus of a secret force that would eventually be identified as the Special Frontier Force (SFF) or Establishment 22 (pronounced Two-Two) and would recruit exiled Tibetans for "behind-enemy-lines activities" if the Chinese were to launch another invasion of India. In effect, the SFF was the first Special Force of Independent India which though not talked about officially, has ample coverage in both domestic media and the internet including ample pictorial coverage.

The creation of the SFF would be dominated by the Indian intelligence community despite using the resources of the Indian Army and setting up base in a small hill town of Chakrata on the old road from Dehra Dun. It would take a few more years before the Indian Army would see the merit in creating a dedicated commando unit that could be used against their traditional rival, Pakistan.

The genesis of what is loosely termed as Army Special Forces did not come through any organisational framework or any dedicated thought of the higher defence management in India. It began, strangely enough, on the initiative of a man who had been passed over for a promotion to the rank of Lieutenant Colonel. This was Major Megh Singh, an infantry officer from the Brigade of Guards who was busy serving as a Grade 2 Staff Officer (Operations) in the Western Army's Headquarters at Shimla. The man heading Western Command was the well-reputed Lieutenant General Harbaksh Singh, a large man whose military acumen was respected by professional soldiers across the country. Singh was heading the largest and the most active Command of the Indian Army that stretched from the Thar deserts of Rajasthan as its Southern-most operational boundaries and the frozen mountains in the North in the state of Jammu and Kashmir.

Singh was also known as a man who was accessible and loved fresh ideas from the rank and file under his command. One fine morning in early 1965, Singh agreed to meet Megh Singh who had come up with an idea to augment operations in the Western Theatre through commando forces. Megh Singh had served in the Patiala State Forces before being assimilated into the Indian Army's 3 Guards. The Army Commander not only agreed to the presentation, he sat through it from beginning to end, as Major Megh Singh detailed what could be achieved through special operations and the advantages of raising a 'Commando Battalion.'

Megh Singh volunteered to raise the Commando unit if he was given the freedom to choose his officers and men. Fascinated by what was presented to him, the Army Commander was quick to see the potential for such a unit as a force multiplier for the troops under his command. Those were days when decisions for new organisations did not require any written sanctions within the Army or the bureaucrats in South Block that housed the Ministry of Defence (MoD) in Delhi.

Singh agreed to the proposal immediately and told Major Megh Singh to start preparations. Nearly two years before Major Megh Singh came to his Army Commander with the proposal to raise a Commando unit, the Indian Army had been going through a process of recovery and modernisation under General J N Chadhury as the Chief of Army Staff. Obsessed with cleansing the ignominy of the defeat of 1962, Gen Chaudhury was busy creating a new army that would be modernised and trained for modern warfare; a decisive shift from a Second World War mind-set to a more modern army that could put up a more credible performance in the wars of the future.

Lt Gen Harbaksh Singh understood the need for change and promised Megh Singh that if the results were successful, he would put in his might behind the creation of a formal and full-fledged Commando Battalion. Maj Megh Singh, described as an earthy man of simple tastes and gritty determination began the task of creating a force that could achieve what normal infantry battalions had not even thought of.

Perhaps Megh Singh was unaware that his initiative was starkly similar to what another young officer had achieved in a similar manner during the Second World War. Captain David Stirling, described by his peers as a rather odd officer, was recovering from a parachuting accident in North Africa during the Second World War when an idea struck him. An obsessive man, Stirling realised that taking this idea to his immediate superiors would not work. He chose to break the chain of command and legend has it that hobbling on crutches he walked into the room of General Niel Ritchie, then serving as the Deputy Commander of the Middle East British forces.

Stirling was looking for the Commander-in-Chief General Claude Auchinleck when he managed to get into General Ritchie's room while being chased by the guards. Stirling managed to convince General Ritchie who took him to meet General Auchinleck. Both convinced the Commander-in-Chief to create a Special Forces unit that could go out on long range desert patrols and hit Field Marshal Erwin Rommel's forces behind their front lines. The new unit that Stirling proposed would be given a misleading name to confuse German intelligence. They called it Special Air Services (SAS), making it sound like a unit created to aid logistics. Megh Singh, who was probably unaware of Stirling's effort, ended up creating a model that was first pioneered by the SAS. Their trajectories would also be starkly similar in the initial years.

Surprisingly, the experience that the Indian Army could have gained with the creation of the SFF in 1963 was never used. Instead, the new Commando unit would start from scratch and learn its ropes on its own. But the commando unit that Megh Singh was putting together was unlike any other unit in the Indian Army. It did not have a name or a designation, without the conventional structures that infantry units were familiar with. The officers and men were all volunteers, each one of them personally selected by Megh Singh. The man was looking for officers and men who were not only better than their peers, they were also built to think unconventionally. Key to the success of this new unit, Megh Singh felt, was the ability to create a cohesion that was never seen in Indian infantry units before this. They would work around the "small team" concepts, which could be infiltrated behind enemy lines to attack their supply lines and inflict attrition far greater in proportion of the size of the attacking force. For the want of a formal name, the officers and men, inspired by this crazy and unconventional leader, christened the unit "Meghdoot Force."

The bulk of the training in physical fitness, tactics and weapon handling would be conducted in the mountains near Udhampur setting gruelling standards that had rarely been attempted in the conventional military. The "small-team" concept was the key as the unit studied maps and the Pakistani military Order of Battle (ORBAT) facing the Western Army Command's theatre of operations to conduct their raids when the war began.

The bulk of the initial manpower came from 3 GUARDS, who were Khemkhani Muslims. The young volunteer officers came from all over;

Subhash Joshi, Arvinder Singh, Hoshiar Singh, Sukhi Mann, JK Sharma, DK Purushe, to name a few. The unit was organised on a six-company basis, akin to the Infantry Battalion.

As the hostilities began in the autumn of 1965, Meghdoot Force was quietly deployed by Lt General Harbaksh Singh across the Line of Control (LC) in Jammu and Kashmir. The unit would send out teams to hit the logistics of the Pakistani Army, infiltrating by first light and exfiltrating within hours of a successful operation. While it did not influence the overall war in any significant manner, it established the success of an idea that Megh Singh had come up with a few months ago. It validated that a small team of specially trained officers and men could inflict severe damage to the enemy's capabilities, far greater in proportion than the size of the force or its fire-power. The nucleus of a special operation had finally arrived.

Needless to say, the results were encouraging enough to move to the creation of a formal Commando unit. Megh Singh, always the leader who believed in leading from the front was awarded the gallantry award of Vir Chakra and promoted to the rank of a Lieutenant Colonel. The "Meghdoot Force" was reorganised into the first Commando Battalion of the Indian Army and was moved to an exclusive location after the war that became their home station for many years. Of the six companies of this Commando Battalion, three companies formed the core when the second Commando Battalion was raised in the deserts in quick succession. Lieutenant Colonel Megh Singh finished his command and moved on to the Assam Rifles and later to the Border Security Force before retiring. His contribution, restricted to a few official papers and records, would be soon forgotten by most people, leaving the man to live a quiet life in Jodhpur, in his home state of Rajasthan, practically forgotten by everyone but the few officers and men who were part of the birth of the Special Forces in the Indian Army.

2

From Meghdoot to 1971 Indo-Pak War and Beyond

The war was over and an appreciative Western Army Commander, Lt Gen Harbaksh Singh kept his promise and had Major Megh Singh promoted. The newly promoted Lt Colonel Megh Singh was asked to "formally christian" the Battalion as the Ninth Battalion of the Parachute Regiment (Commando). As word spread about this new special unit, spirited young officers from across the Indian Army rushed to volunteer and serve.

Youngsters like Lieutenant N S Rathore (ex 4 Para) and Lt Tej Pathak, (ex 1 Para) were the young officers who signed up for this new battalion. At that point of time, officers from the Parachute Battalions were not expected to undergo any special training to join the Commando Battalion but Lt Col Megh Singh was clear that the junior leadership had to be better than their counterparts in the conventional army. He began to send his NCOs and JCOs in batches to the Infantry School in MHOW, an old cantonment town, MHOW being the acronym for Military Headquarters of War established during World War II.

Located in the heart of India, in the erstwhile Central Provinces that would become Madhya Pradesh in independent India, MHOW was a cantonment left behind by the British. It was here that Lt Col Megh Singh started sending his men for special training in sniper firing and other support weapons. These were the only special skills available under formal army training institutes to the fledgling unit and the Company Commanders, known as Group Commanders in those days, had to devise their individual training methods.

Some groups would be out at night in and around Gwalior, where the unit was being raised, doing speed marches and mock raids at night. The emphasis was on physical fitness, navigation, and learning any skill that they had read about in foreign journals and magazines that managed to find their way into the Officer's Mess.

Lt Col Megh Singh got a new Second-in-Command in Major Bhawani Singh, an officer who belonged to the royalty of Rajasthan and was keen to take the expertise developing in the unit to a separate unit that would specialise in operations in the desert albeit rumour has it that the relationship between Lt Col Megh Singh and his Second in Command was not on the best of terms.

In July 1967, a decision was finally taken by Army Headqurers to establish two battalions with different theatre level special skills. The original unit, Raised by Lt Col Megh Singh would head off to Udhampur to specialise in conducting operations in the mountains, while the others would head off to a location in Rajasthan and specialise in commando raids in the deserts.

On a blazing afternoon on the Gwalior airfield, six companies of the unit were asked to stand in file formation and then split up into two units of three companies each. The original unit, already christened as 9 Para (Commando) would now have a sister battalion in 10 Para (Commando) under Lt Col N S Utthaya. Naturally, Maj Bhawani Singh from Rajasthan would be the Second-in-Command of the new unit and take over once Lt Col Utthaya had moved out. Most of the troops from Rajasthan and those belonging to the Jat community went to 10 Para (Commando) while those from the Dogra caste stayed back in 9 Para (Commando). Lt Col Utthaya would take his men to Nasirabad and start training his men in the ways of the desert before eventually moving further West to Jodhpur.

Lt Col Megh Singh would return to Udhampur, close to the Corps Headquarters, which would oversee most operations in that sector. By December 1971, both units were in a position to operate against specific tasks if war revisited the Sub Continent.

Faced with a massive exodus of refugees from East Pakistan, Prime Minister, Indira Gandhi was on desperate tour of the West trying to drum up support for India and the refugees fleeing Pakistani repression. In East

Pakistan, the leader of the Bengali political party, the Awami League had gained an absolute majority and was in a position to form a government in both East and West Pakistan. This was unacceptable to the Punjabi-dominated West Pakistan and a third war with India was just a matter of time.

By this time 9 Para (Commando) was already well entrenched in the North and was training for specific tasks. While Alpha and Bravo Teams were sent off to intended deployment areas, Charlie Team would play a decisive role in one of the first commando raids across the line of control.

Alpha Team was tasked along with an Infantry Battalion and a troop from an Armoured Regiment to defend Poonch under command of Major MM Cariappa. Alpha Team, despite being part of a commando unit was tasked for a conventional role, a mistake that would be repeated in later wars including with the same unit. One of the youngest officers in the unit, Lieutenant Hardev Singh Lidder was posted in Commando Wing, Belgaum in Karnataka when the war broke out. Lidder immediately took leave and pushed off to Jammu and Kashmir to join his unit and joined Alpha Team in defending Poonch after three days of grueling travel. It is believed that a Pakistani commanding officer was captured by the commandos during one of the battles fought in the area.

Bravo Team was tasked by an Infantry Division to raid multiple ferries in enemy area and hold them till the war was over. Their plan was to inflict attrition on the enemy in the same manner the Meghdoot Force had done in 1965. Led by Captain N S Rathore and his second in Command, Lieutenant Tej Pathak, the team was initially tasked to raid these multiple ferries on the same night. Captain Rathore went up to the Divisional Commander to request a change, worried that a failure to achieve the task would give his unit a bad name. The Divisional Commander saw merit in Captain Rathore's argument that each ferry had to be tackled separately, one after the other and he readily agreed to the change. The Corps Commander, Lt Gen Sartaj Singh, known as a man who would not budge after a decision was taken was kept in the dark about the change.

With sketchy intelligence support and warnings not to fire at the enemy without provocation Capt. Rathore launched his team after 0000 hours. Capt

Rathore had a team of 150 men and five officers. Navigating through the elephant grass the team suddenly came across Pakistani troops positioned near the ferries. At first they mistook the Indian commandos as reinforcements from Sialkot and immediately announced that the "Kafirs had attacked." Capt Rathore could have capitalized on the confusion but instead chose to fire at the Pakistanis who immediately returned fire and ran away.

A squadron of Indian tanks had been kept in readiness if the commandos ran into any Pakistani armour but for some inexplicable reason they had begun withdrawing as soon as war was declared. Capt Rathore and his team arrived at the target ferry and after a brief firefight captured it. He would repeat the same feat a few hours later at the second ferry and hold both of them till the end of the war.

Further North Charlie Team was preparing for a major raid across the Poonch River into a Pakistani artillery unit that was harassing Indian positions. Mandhol was a sleepy village across the Line of Control in the Mendhar sector. Six Chinese 122 mm guns had been placed there and were proving to be a major problem for the Indian positions. Originally an Infantry Battalion was tasked to launch an attack on Duruchian but the plan was dropped at the last minute. However, Charlie Team would continue with its plan to take on the guns at Mandhol.

Like the other teams, Charlie Team had its share of young and energetic officers like Lt Roshan Lal and Lt Ashok Taskar, and led by Major O P Malhotra, a veteran of the 1965 war. Accompanied by Capt D Tyagi, an artillery officer, Charlie Team infiltrated into Pakistan Occupied Kashmir at 000 hours. After entering the village they found it deserted with an old man left behind as a care-taker. The guns had been dug into positions around the village and would fire daily at Indian positions leading to major attrition to the defences in and around Poonch Sector.

Charlie Team crept up to the guards and launched their assault at 0200 hours, and swiftly went about blowing up the guns. By 0300 hours the Team had accomplished its task and with a few casualties began the long trek back home. The Mandhol raid was a unique action that Indian troops had not undertaken before. It came closest to the classical commando tactics of raiding a position of great importance to the enemy and causing attrition that would provide gains indirectly proportional to the size of the raiding party. For a young unit, the success at Mandhol would become a legend that would establish 9 Para (Commando) in coming years.

Down south, Lt Col Bhawani Singh was busy preparing 10 Para (Commando) for a raid across the international border driving through the deserts in Pakistan opposite Barmer. His plan was to raid the headquarters of the Pakistani rangers in Chachro, a town in Sind, Pakistan, nearly 80 km from the international border. Riding specially-outfitted jeeps resembling Capt David Stirling's SAS patrols during the Second World War, a teams from 10 Para (Commando) led by the Bhawani Singh would ride into town and take out the Rangers Headquarters. Infiltrating in the early hours of December 6, Lt Col Bhawani Singh and his raiders drove to Chachro and attacked the Headquarters taking the Pakistani troops completely by surprise. A similar dash across the international border almost proved to be a major problem when nervous Indian infantry units opened fire on the returning raiders, mistaking them to be Pakistani troops.

While 9 and 10 Para (Commando) had established themselves as "special units" in war, they were still commando units, unlike their counterparts in the West who had moved on. The British SAS, already an established Special Forces' unit, had played a decisive role in the counter insurgency operations in erstwhile Malaya and had also been employed by the UK Foreign Office in the Middle East. SAS officers and men would play a major role in Yemen, Oman and a string of other Arab countries in a strategic role to win goodwill for Great Britain's strategic interests. Modern forms of terrorism had also begun to pose numerous challenges for Special Forces units across the world and a few years later would see the Israeli Sayeret Matkal fly over 2000 km across Africa to rescue their citizens from Entebee, the capital of the African nation of Uganda.

For the Indian commandos, the decade would be one of relative peace with most of them spending time in training. Till 1978, the two units would continue to prepare for future wars till Army Headquarter decided to convert another battalion to a commando unit. The war in 1971 had seen both the units distinguish themselves. However, Army Headquarters felt that a third commando battalion was needed to serve as a strategic reserve that would serve directly under it in times of war or conflict. Little did anyone know that with insurgency slowly emerging in the state of Punjab, the commandos would soon have to be ready for a new kind of conflict. The next major intervention for the commando units was still a few years away and Army Headquarters got down to the task of converting an existing Parachute Battalion into a modern commando unit. For the task they chose the oldest battalion of the Indian Army (raised in 1761), a unit that had changed numerous designation over decades and from a later avatar as First Battalion of the Punjab Regiment that had been converted into the First Battalion of the Parachute Regiment. 1 Para had distinguished itself in both the wars and one of its officers, the legendary Ranjit Singh Dayal, a young army Major in 1965 who led his team into battle to capture the Haji Pir Pass, himself winning the Mahavir Chakra for gallantry. For the men and officer of 1Para, it was time to convert once again and chart new territory.

3

Conversion and Counter-Terrorism

During the Army Commander's Conference of March 1977, a decision was taken to raise a unit that would resemble the British SAS. The idea was to create a true Special Forces unit, unlike the existing Para-Commando units that already existed with specialisation in conducting operations in the mountains, plains and deserts.

To achieve this, Army HQ decided to take 9 Para (Commando) and convert it into an "Experimental Commando Wing" that could conduct tasks similar to the SAS and concurrently convert another Parachute Battalion to a Commando Battalion for operations in mountains. The SAS had evolved into a single battalion with multiple skills. Called the 22 SAS, the unit was based out of Hereford from where the squadrons (equivalent to the Indian infantry company) would be rotated to accomplish different tasks. They would be employed in a strategic role or in a counter-terrorism role, depending on the requirement. A directive was issued in 1978 to raise an Experimental Commando Wing and undertake trials with this organisation.

However, Lt Gen Inder Singh Gill (recipient of a Military Cross) as the Colonel of the Parachute Regiment held a different view. Few outside the military knew that Lt Gen Gill was one of the few officers who had been part of several commando raids and "behind-enemy-lines" activities during the Second World War and had even successfully escaped from a prisoners of war camp though shot in the neck. Born to a Sikh father and a Scottish mother, Lt Gen Gill had served in the British Army as an undercover operative in Nazi-occupied Greece. He was part of *Operation Harling* a secret plan executed by Allied Headquarters to help create resistance movements in the occupied territories and lead them in guerilla raids. They would go around blowing up bridges and enemy supply lines to cripple them from within while the conventional war raged on hundreds of miles away. Lt Gen Gill, a former Commanding Officer of 1 Para had a sound argument.

Lt Gen Gill argued that if 9 Para (Commando) was to be converted then it would leave the Indian army with fewer options to employ commando units in the event of a war in immediate future and another Parachute Battalion under conversion would also not be available in such contingency. Instead he proposed that the oldest Parachute Battalion be converted into an Experimental Commando Wing on the lines of the SAS with a War Establishment (WE) profile that was similar to the British Special Forces. The conversion commenced in 1979 and a trial report was submitted in May 1982.

But the conversion did not come without its attendant problems. Once again, the experience gained in raising the SFF and 9 and 10 Para (Commando) was set aside, or probably forgotten by a military hierarchy that would not think beyond tactical level employment during conventional wars. Rather than raising fresh Special Forces units, existing conventional infantry and parachute units kept on getting converted to Special Forces role over the decades. After, the three Commando Battalions, more than a decade later another conventional unit was converted into a Special Forces role, and thereafter repeated again and again. The standards of selection, probation and manpower policies continued to be stuck between pragmatism and tradition. Some officers chose to leave the battalion. Those who chose to stay back had to strangely undergo a split probation; both at the Commando unit they were to join as well as at the Parachute Regimental Centre in the southern city of Bangalore,

Part of the problem was a cultural construct that was steeped in the army's colonial legacy. Having been created by a colonial power, the Indian Army continued to follow the legacy of the British in its traditions. For an army that swore by traditions, it was a challenge to move away from age-old concepts of class and caste-composition of battalions, command and control structures and conventions that had remained valid for decades. To suddenly move away from that and embrace modern concepts of war like developing a special operations capability while retaining its traditions was a challenge. That, probably, led to a gap between the original thought to achieve a capability similar to the SAS, and what was achieved in the conversion of 1 Para into an Experimental Commando Wing.

To understand this, the evolution of the SAS into a modern and feared Special Forces unit offers an interesting context. The British Military respected its traditions as much as any other army. But the Second World War, thrust upon them with the might of German blitzkrieg and the battle for Britain forced it to look for unconventional solutions when faced with a numerically and technologically superior force. Born in the blitzkrieg of North Africa, the SAS offered a similar unconventional solution taking on to Erwin Rommel's Panzer divisions. What Captain David Stirling offered was elegant in its simplicity and daring in concept. He wanted to create an unconventional force that would slip in behind enemy lines and cause tremendous damage to the fuel, oil and lubricant supply lines of the Panzer divisions. His commandos never confronted or attacked the powerful tanks. Instead, he found their vulnerability in their logistic supplies and knocked them out during numerous Long Range Desert Patrols (LRDPs).

In fact, Captain Stirlling's efforts were aided by another unconventional British officer who had helped the Jewish underground create a similar unconventional narrative in Palestine. His name was Brigadier Orde Wingate, a man who was obsessed with the Bible, could speak Hebrew fluently and made friends easily.

While Captain Stirling's commandos raided the supply lines of the Germans in North Africa, Brigadier Wingate was setting off for India with plans to put his unconventional ideas into motion on a grander scale. His plan was to create a force that would work closely with the Royal Air Force and infiltrate behind the Japanese defences and cause them incalculable harm. After training extensively in Gwalior, the troops, now christened as Chindits or Wingate's Raiders were infiltrated into Japanese held Burma (Myanmar). Promoted to a Major General, Wingate had to fight numerous battles with his hierarchy to convince them about the efficacy of the Chindit operations. While declassified interrogation reports of the Japanese Divisional Commanders is now comprehensive proof of the success of the Chindit operations, a chicanery by one of Major General Wingate's enemies left a distorted view of the Chindit operations for generations of Indian army officers passing out of the Indian Military Academy (IMA).

Every Gentleman Cadet at the Indian Military Academy (IMA) amongst the various books recommended, gets to read Field Marshal William Slim's account of the war in South Asia recorded in his memoirs *Defeat Into Victory*. The book maintains an unflattering account of the Chindit operations and has been assumed to be fact for generations of officers since then. But the declassification of the interrogation reports of the Japanese Divisional Commanders has now conclusively proved that the Chindits succeeded. A further investigation has revealed that Field Marshal Slim was sending his chapters to another officer who had served under him during the War. The man was Major General Richard Kirby, who would also author the official history of the British Army's operations in South East Asia during the Second World War. What was not known to many at that time was the Major General Kirby was one of the bitterest critics of Major General Wingate.

For months Major General Wingate had fought a bitter battle with Kirby for holding up his supplies and operational requirements. Major General Wingate was known as a man who could not suffer fools and had many bruising sessions with Kirby. After Major General Wingate's untimely death in an air crash, it was left to Kirby to change the historical perspectives on the real success of the Chindits. That legacy would continue to play a role as the Indian Army continued to hold on to the legacies of the British. Ironically, the SAS, disbanded after the Second World War, would find a new role to justify its existence in counter-insurgency operations in Malaya, to be led by a former Chindit officer. It was the perfect example of Captain David Stirling and Major Orde Wingate's legacies coming together to give a new lease of life to what is perhaps one of the finest Special Forces units in the world today.

It was in this context that Army Headquarters started the conversion of 1 Para without looking at either equipping them differently or even tasking them in unconventional roles. Luckily, the conversion was done under a dynamic Commanding Officer, Col Avtar Cheema who had scaled Mount Everest as a young Captain, which happened to be his very first attempt in scaling any mountain peak. The Second-in-Command in Lt Col Sukhi Mann was an equally dynamic officer. But the conversion was done carrying the same 7.62 mm Self Loading Rifles and the unpredictable Sterling sub-machine guns that were the successors of the Second World War "sten-guns". While special weapons and equipment came much later, for many years the army hierarchy was locked in mundane debates like whether to permit personnel of these special units to operate in civil attire at all.

The unit converted to the new role with greater emphasis on physical fitness and small team actions. However, the orthodox hierarchy failed to let it graduate to what the original vision had been; to emulate the SAS. Interestingly, an American Special Forces officer, Major Charles Beckwith, who had served with the SAS returned home to create a unit on similar lines. His experiment succeeded and gave birth to Operational Detachment Delta, popularly known as Delta Force.

Col Avtar Cheema was fully aware that no amount of training could compensate for actual combat experience. So, he invited General KV Krishan Rao, then Chief of Army Staff to witness an exercise of the newly converted unit and even strapped the Chief right next to an open aircraft exit door thousands of feet above ground level while commandos with full combat loads parachuted and sky dived for exercise tasks. This was an experience not witnessed earlier by any Service Chief. Once on ground, Avtar Cheema told Gen Krishna Rao that he would like to send assault teams of his unit by rotation for counter insurgency in the North East, to which the Chief readily agreed. Consequently, assault teams from Para (Commando) battalions located in peace stations started going for counter-insurgency operations to the North East; Nagaland, Manipur and Mizoram and acquitted themselves well.

Post the conversion-cum-raising of the Experimental Commando Wing, the two erstwhile Para Commando Battalions also converted to the same War Establishment (WE) with three assault teams each as their teeth.

But the first major action where the commandos would be called upon to operate in dense built up area was brewing in Punjab where Sikh militants were beginning to assert themselves. The revered and holy Golden Temple had become their base, stocked with weapons and turned into a fortress of some sorts. After negotiations between the government and the militant leadership failed, then Prime Minister, Indira Gandhi ordered the Indian Army to flush them out.

Perhaps, the military hierarchy's lack of familiarity with special operations added to the complexity of the proposed operation aside from practically nil intelligence on the holed up militants. *Operation Blue Star,* as the operation came to be called, was a complex operation, with several disparate elements being brought in to deal with different aspects of a complex puzzle. A unit minus of Para Commandos became one such element that was brought in to deal with only one aspect of the problem and not attack it as a whole. This led to a situation where every unit brought in for flushing out the militants it seems, was on different missions.

The Para Commandos were moved with only few hours warning and reached the area just 48 hours prior to the operation. For the commandos, brought in hastily to augment the action, the intelligence support was abysmally lacking. Carrying their antiquated weapons, the commandos began making preparations for their role that would be recorded as *Operation Metal* by the unit. Even as the commandos were making their plans, a team from the Special Group that worked under the Cabinet Secretariat, part of the SFF, arrived with a different plan.

Special operations succeed when Special Forces are given the task, leaving the methods how to achieve the task to the specialists. Unfortunately, here they were directed that they could enter the complex only through the regular entry, covered by heavily armed militants, denied the use of armoured personnel carriers and forbidden from even using their 84 mm rocket launcherss. The request to use explosives to breach and make their own entry points was declined firmly. The commandos didn't even get a chance to rehearse their attack. To top this, the night chosen for the operation was a full moon night which would reflect the moonlight in the marble flooring denying them even minimal cover of darkness.

The Commandos had to enter from the public entry point to the Holy

Temple that was well covered by the militants. Forced into an impossible situation, the commandos went into action with nothing but guts and their antiquated firepower. They virtually walked into a designated killing zone. The Commanding Officer later briefed the media that they were sent in "with their hands tied behind their backs". As the operation continued, the commandos continued to take in fire before they could breach the target.

Operation Blue Star and/or *Operation Metal* was conducted without any understanding of special operations. There were no hostages so that did not put the army leadership under any kind of pressure to launch commando troops in frontal assault in such fashion. They could have planned the whole action as a special operation and left it to the commandos drawing from all the three existing battalions. However that was not to happen and the military commanders tried to execute a complex mission that failed to make a significant impact until heavy casualties began to rise.

The operation was flawed from the very beginning and the employment of the commandos was ill-conceived. Overall, the military relied only on brute force, fighting on exactly opposite to the age old teachings documented in the army manual on 'Fighting in Built Up Areas' and suffering avoidable casualties. This would continue to haunt the commandos in the wars they would fight in the future as senior military commanders with little or no understanding of special operations continued to throw them into impossible situations. Worse, they would be employed as conventional infantry units despite the fact that they had been converted to operate on levels other than the conventional.

While *Operation Blue Star* was an unmitigated disaster for the commandos their counterparts in other parts of the country were busy with missions of their own. New Delhi had begun receiving reports of Pakistan encouraging treks and expeditions to glaciated regions north of NJ 9842, the last point on a contentious map that had been hastily drawn in the aftermath of the Partition. The Cease Fire Line between the Indian and Pakistani troops drawn during the 1949 Karachi Agreement (after the conflict in 1947-48) and later even during the 1972 Shimla Agreement had never looked at this strategic frozen area (especially the Saltoro Range) and the area had remained quiet but for occasional Pakistani claims that the land belonged to them. A chance Indian patrol saw Pakistani troops camped on the Western slopes of the Saltoro range, preparing to climb up (Gen Musharraf later acknowledged in his autobiography "In the Line of Fire" that Pakistan was planning to establish an infantry battalion on the Saltoro Range). Worried by these signs and signals emanating from the West, the Indian Army was tasked with establishing a presence on the Saltoro Ridge by launching *Operation Meghdoot*. Three units were chosen to lead the initial helicopter borne assault.

A year after *Operation Blue Star*, the mountain commando unit was called in to check out some violations on the Line of Control as a part of *Operation Meghdoot*. The farthest Indian Army post on the LC, known as 'Gulab' Post was a stretch from Turtuk near Kargil. After 'Gulab' Post, the LC ran to NJ 9842 that and from there the Saltoro Range lay northwards. However intelligence reports of Pakistani violations had come in area that lay between Turtuk and Gulab Post in October 1985.

A team from the mountain commandos was launched to scout around the area known as the Laoching *Nala* that almost stretched to around 15 kilometres. A surveillance observation post was set up by the para-commandos and some sporadic firing was exchanged with the Pakistanis. Nearly 48 hours later a four-man patrol was launched to scout the area from another spur in the mountains. Unfortunately, three of the commandos were killed when the Pakistanis retaliated with heavy fire and only the officer leading the patrol managed to return. This would lead to several questions about the efficacy of the operation but the fact was that the commandos were undertaking reconnaissance when the numerically superior Pakistanis troops infiltrating in the area chanced upon them.

Between 1971 and 1984, it had been a relatively quiet time for the Indian commandos barring the actions in the Golden Temple and general area of Siachen. In the aftermath of the tragic assassination of the Prime Minister Mrs Indira Gandhi, the mountain commandos and the Special Group had been rushed to New Delhi to provide security for the new Prime Minister. Barring these sporadic incidents, it was a relatively quiet time for the commandos other than counter insurgency operations in the North East from early 1980s.

The conversion of a conventional Parachute battalion into a commando role had delivered mixed results since their originally intended roles akin to the SAS were forgotten post conversion. Their future was uncertain and no major road map had been set out for them. Once again they would be left to improvise on their own, while Army Headquarters continued to function on a day to day basis, putting out fires when called upon to do so.

Things would have gone on like this but for a chance trip to the United Kingdom by a young Lieutenant Colonel from the Gorkhas who held a fascination for the Special Forces. His visit to the UK would prove to be decisive moment for the future of the para-commandos.

4

An Idea Whose Time Had Come

Every cause needs a champion. If the Indian Special Forces needed one, they found it in Lieutenant Colonel Rustom K Nanavatty, an officer who had been commissioned into the Gorkha Regiment. Lt Col Nanavatty had been selected to go to the United Kingdom for a little-known posting that had survived from the pre-independence days.

The Indian Army Liaison Officer (IALO) in the School of Infantry, Warminster, UK was a little known position that had great value for mid-level career soldiers. It offered them a chance to visit and interact with the British Armed Forces at a professional level and draw lessons that could be used to benefit the Indian Army back home. The post actually provided a window to get an insight into modern armies of the West. For most Indian Army officers it was chance to take a break from the humdrum of daily soldiering and stay in the UK for a year or two and visit the rest of Europe on a tight budget, in line with his parting military briefing to "go have a ball". But unlike most others, Lt Col Nanvatty was busy trying to get a better understanding of how the SAS worked and operated.

As the IALO, Lt Col Nanvatty was allowed to access any material up to the level of 'Confidential" classification and visit any training establishment or formation barring two formations. The restriction on the first formation was the hardest for him to overcome: he was not allowed to visit the SAS Headquarters at Hereford. The British Government also barred him from visiting any area in Northern Ireland since it was still reeling from terrorism. A foreign armed forces officer being sent home as a casualty would be deeply embarrassing, the British authorities informed him.

However, after receiving several requests from a persistent Lt Col Nanavatty, they allowed him to visit the HQ of the Director, SAS (HQ DSAS) at the Duke of York HQ, Chelsea on a special invitation. He had been warned by the hosts that the visit would be cursory at best and the interaction would be limited for security reasons. On February 22, 1985, Lt Col Nanavatty finally arrived at the HQ DSAS to discuss several issues that included:

▸ The role of the Directorate,

▸ Selection and training of the personnel,

▸ The career structure of the officers and soldiers,

▸ Command and control and

▸ Assistance to foreign governments.

Through the day Lt Col Nanvatty was given two presentations and several informal discussions. A film classified as 'Secret" on the role, selection and the various individual skills that they possessed was shown to him. This was followed by a film demonstrating four of the seven methods of countering a hostage situation that the SAS practiced in those days.

Impressed with what he had seen Lt Col Nanvatty wrote back a detailed five-page report to Army Headquarters along with an assurance from the SAS that they would be "delighted" to host an Indian Army Brigadier since the British Ministry of Defence had cleared it.

Lt Col Nanavtty's report and his enthusiasm would spark off a churning in New Delhi that was finally beginning to appreciate that there was more to Special Forces than the commandos who were available in the three Para Commando Battalions. Major General Bipin C Joshi, who would later take over as the Chief of Army Staff was posted as the Additional Director General of the Perspective Planning Directorate. Gen Joshi saw merit in the ideas that Lt Col Nanvatty was propounding and after taking over the powerful Military Operations Directorate as DGMO, Lt Gen Joshi agreed that it was time to do a reappraisal of the Para Commandos and prepare a road map that would help them emerge as Special Forces.

After convincing Army HQ, Lt Gen Joshi picked up Brigadier N Bahri, an officer who had been commissioned into the Brigade of Guards, just like Lt Col Megh Singh, who had formed the Meghdoot Force. Brig Bahri was considered as a very balanced, educated and professional soldier. He had an open mind, questioned everything and suffered from no biases. These proved to be an advantage for the three-man committee that had been set up to conduct a seminal study and propose a reorganisation of the Para Commandos. Naturally, Army HQ picked up Lt Col Nanavatty and another veteran Para Commandos, Lt Col Sukhi Mann as the other members of the committee.

The Committee was given four terms of reference:

▸ Organisation

▸ Employment

▸ Manpower

▸ Equipment

They were not asked to look at the force levels since the three existing para-commando battalions were considered adequate to deal with all exigencies. The Committee began work immediately and prepared a comprehensive report that sought to reorient the Para Commandos into true Special Forces.

The Committee suggested that to achieve this road map, the existing Para Commandos had to be separated from the Parachute Regiment and be reorganised into a separate Special Forces Regiment. This was to ensure that the military hierarchy recognised the two significantly different roles played by these components forcibly yoked together for decades.

When the first Para (Commando) battalion was raised, they chose to house it in the Parachute Regiment mainly to facilitate delivery of commandos to their objectives. But fundamentally the Para Commandos (or Special Forces) are different from their brethren in the Parachute units. While it is without doubt that the Paratrooper is justifiably considered among elite troops, he is actually not very different from the conventional infantryman albeit with airborne capability. After being delivered to the objective by airborne means, he assumes the role of an infantryman and is relieved with the ground troops who are several hours behind. The paratroopers are expected to play various roles as pioneers or to capture some strategic objectives and hold it till the relieving infantry arrives. But they are not Special Forces and their roles have no similarities. But the controversy continued to rage on even despite the Perspective Planning Directorate, Army Headquarters penning down in 1999, *"By their very nomenclature, the Parachute (Special Forces) Battalions are unique. Therefore, there is a need to de-link and distinguish them from other similar organisations. While some of the tasks that they may perform may be similar to those of regular Infantry / Parachute Battalions, the similarity ends here. Unless this uniqueness is accepted and enforced in all aspects of organisation, equipment, training and administration of these units, they may not achieve their potential"*.

This was a fact recognised by the Brig Bahri Committee which proposed that a separate Special Forces Regiment be raised with its three battalions on lines of the SAS Regiment.

The Committee also felt that the current structure of the Para Commando units being dedicated to Army Commands was not the best way to progress. In the UK, the SAS was a centralised formation with a Brigadier from the SAS serving as a Director of the Regiment with direct access to the Prime Minister's Office. This offered several advantages to the military hierarchy as well as the Special Forces. The military had an expert who could offer sound advice to the political leadership about the capabilities and employment of the Special Forces in situations that demanded their intervention. It also gave the military leadership the opportunity to seek sound professional opinion during exigencies, giving the military hierarchy to work with the political leadership at various levels.

To achieve better command and control, the Committee suggested that a separate cell be created in the Military Operations Directorate that would report directly to the Vice Chief of Army Staff through the Director General of Military Operations. While the individual Special Forces' Battalions would continue to be deployed in the various Army Commands and report to the formation commander for administration and discipline issues, operationally they would have a more centralised command and control system. This helped them achieve a similar kind of centralised approach that the UK and the United States of America had already adopted. In fact, the United States had gone a step further and created a Congressional mandated structure, known as the Nunn-Cohen Amendment to the Goldwater-Nichols Reorganisation of the Pentagon Act creating a separate Special Operations Command (SOCOM).

The Committee felt that since the centralised approach of the other militaries had already proved to be a success there was no need to reinvent the wheel. Centred around this basic recommendation were the other recommendations for manpower selection, training, equipping and tasking.

An elaborate structure was recommended that there was a need to:

- ▸ Modify the tenure system

- ▸ Create an All India class composition

- ▸ Bring in trained soldiers from all Arms and Services

- ▸ Ensure no Extra Regimental Employment (ERE)

- ▸ Improve the incentives for the men selected into the Special Forces

The Committee also detailed separate policies for officers and men to ensure that a system was created where the best volunteers could be screened and inducted into the battalions. Every volunteer was now expected to undergo a probation that would last for almost a year along with continuous training in acquiring special commando skills. The tour of duty for volunteers from units other than the Parachute Regiment would be for four years.

For the potential officers, the Committee suggested that a special briefing be held regularly so that they could be interviewed, advised on how to prepare for the grueling months of selection tests and also get special advice on how to plan their careers after a stint with the Special Forces. The Committee felt that these officers could either continue in the Special Forces or go and proliferate their specialised skills that would benefit the other Arms and Services. Interestingly, the Committee recommended that additional career weightage should be given to all the officers and men who had served with the Special Forces. A similar schedule was worked out for training of the officers and men.

The Committee had already recommended a structure to impart this special training. A Special Forces Training Wing was to be created, to be headed by a Lieutenant Colonel. This establishment would ensure the imparting of special skills like sniper training, surveillance, combat free-fall training and other Special Forces skills. This establishment would also become the repository of the collective wisdom of the Special Forces units to create an institutionalised memory that would benefit the organisation in the future.

The Committee also felt that certain concrete steps needed to be taken to create a true Special Forces capability. This would be structured under a two-tier system. The first tier would be the Para Commando units for the conduct of theatre specific tactical operations. To achieve this it was proposed that:

▸ Redefine role, missions and capabilities of Para Commando Battalions.

▸ Review existing structure.

▸ Convert only one assault troop per unit into a military combat free fall troop.

▸ Concentrate all underwater diving skills into an independent amphibious team.

▸ Expand Para Commando Cell at Army Headquarters with operations and intelligence functions to remain with the Military Operations Directorate. Assistance and technical control was to be exercised by a Colonel designated as Director Commando Cell.

▸ Institute a Para Commando Training Wing.

▸ Raise one more unit/sub unit to meet the operational requirement of various commands.

▸ Ensure 9 and 10 Para Commando are brought on par with 1 Para Commando where organisation, weapons and equipment and employment are concerned.

- ▸ Formulate revised individual training policies.

- ▸ Allot available resources to Headquarters Commands for the conduct of theatre related tactical military special operations tasks based on operational requirements.

In the second tier, the committee felt that a Special Group be created for the conduct of a national level and Army Headquarters related strategic military special operations. Army Headquarters would exercise command and control of these units through a Special Forces Headquarters. This would need several steps to be undertaken, which were:

- ▸ Define the role, missions and capabilities of this Special Group

- ▸ Takeover the current responsibilities from the SFF

- ▸ Establish a Special Forces Headquarter

- ▸ Establish a Special Forces Training Wing

- ▸ Ensure that there was proper interface between the Para Commandos and the Special Group. The Para-Commandos would also ensure a regular feed to the Special Group.

On paper, the recommendations of the Committee were a revolutionary step for an institution that had thrived on tradition and convention. If the military hierarchy could be convinced to accept these recommendations, then the second phase of these "reforms" could begin right away. These were exciting times for the men in the Para Commando units. They were aware of the Committee working on a major revamp and they were looking forward to the implementation of the recommendations at the earliest.

But by the end of 1988, when the Committee began to wound up its work, trouble was brewing in Sri Lanka. The indigenous Tamils in the island-country were busy regrouping themselves into a guerilla force to take on the Sinahala majority government. With the Tamil dominated northern regions of Sri Lanka cut off, the situation was threatening to become a major concern for the international community. New Delhi was concerned and an accord signed between India and Sri Lanka led to the creation of an Indian Peace Keeping Force (IPKF) overnight.

Once again, the Indian Army was being called to serve India's national interests in a major conflict. Few in the military knew what the political or strategic objectives were aside from conduct of elections. But for those in the fledgling Special Forces community, this was an opportunity to validate their sills in battle. All the three Para Commando units would be inducted into the island through 1987-1988 and witness some of the bloodiest counter insurgency battles in the history of the Indian armed forces. For the Indian Special Forces community Sri Lanka would be a decisive footnote in their evolution. The tactics and the skills they would be forced to learn would come to their aid in the battles they would fight in the future. For many of them, Sri Lanka would soon become an essential rite of passage. It would also present an opportunity for the three Para Commando units to become Special Forces units.

5

Op 'Pawan' - The IPKF Experience

It was October 1987 when the call came to the Mountain Para Commando unit post-lunch telling them that they had to leave at first light for a place called Palaly in Sri Lanka. None of them had ever heard of the place and it took a few minutes of intense searching on the map to figure out where they were headed. Ironically, even a few years later when then Naval Chief, Admiral Vishnu Bhagwat sent a paper to the Ministry of Defence (MoD) for integration of the Special Forces of the three Services, the Army's Military Operations Directorate view was that they did not visualise trans-border employment of the their Special Forces !

Ever since the operations near the Saltoro Range West of Siachen Glacier area in 1984-1985, things had been quiet for the mountain Para Commandos. In fact a few hours before the call came, a few of the officers and men were all set to leave for Mumbai the next day. A polo match was being organised and a Para Commando team was supposed to make sky diving jumps as a special show at the Mahalaxmi grounds in Mumbai. For the participating men this was an excellent opportunity to get out the boondocks of Udhampur and see the city of Mumbai and be treated as VIPs for a few days.

But the call changed all that and the battalion went into frenetic activity as the assault teams began to pull their gear together. All the officers were rolled into the sand model briefing room and the Commanding Officer briefed them with the sketchy details that had been passed on to him from Army HQ. The next day at 0600 hours, the first of the Antonov 32s landed to pick up the men and fly via Nagpur and Chennai before landing at the Sri Lankan military airbase at Palaly in the evening.

A year before the call, the unit had been practicing grueling exercises designed by Col Tej Pathak, its new Commanding Officer. Col Pathak was a veteran of the first commando raid across the Line of Control in the Poonch sector during the 1971 war. He had been an integral part of the battalion and was steeped in the Special Forces tradition. He had a fair inkling that the unit needed to change its orientation if it had to survive future wars and he put together a punishing schedule for his commandos. There would be night marches and camps in high altitude areas where they would be expected to maintain strict time lines. Col Pathak knew that it was important to shape the unit from a mere Commando unit into a true Special Forces unit. Having spent time reading through literature on the SAS and the US Special Forces, he began to formulate training methods that were unique. He would start working on the firing skills of his men and also their ability to navigate. The idea was to create small teams with excellent and varied skills that could be used when inserted behind enemy lines for special operations.

Slippages would be looked down upon and there was a growing anger within the unit at the tough methods being used to train them. But a few weeks into Sri Lanka most of the officers and men began to realise that Col Pathak's tactics, though brutal and unforgiving, would prove to be of immense value in the killing fields of Sri Lanka. When the conflict in Sri Lanka would start, the hard training would be responsible for honing raw material that was ripe to be moulded into a good Special Forces operator.

Most of the Para Commando teams went in with their antiquated weapons – the 7.62 mm Self-Loading Rifle (SLR) and the unreliable carbine. When they came in contact with the LTTE (Liberation Tigers of Tamil Elam) guerrillas, they were confronted with the efficacy of the AK-47 assault rifle, the preferred weapon of choice of the enemy and for that matter most terrorists and insurgents around the world even today. In fact, as operations began to roll out, any captured AK-47 was considered as the best prize they could win. The captured AK-47s became a huge favourites and since the rifle used the same 7.62 mm ammunition in the SLRs, they decided to hang onto the captured weapons. The weapons carried by the Para

Commandos are an essential adjunct to understanding the kind of challenges that they faced while conducting operations in Sri Lanka. A desperate Defence, Research & Development Organisation (DRDO) unable to deploy the under development INSAS rifle (the development and fielding of which took an incredible 15 years), rushed in a fully automatic version of the 7.62 mm SLR. The Para Commandos christened it as One-Charlie, a weapon that was so unreliable that every burst would force the weapon upwards. The joke among the troops was that the weapon had been designed for anti-aircraft fire, rather than close quarter combat. Many months into the IPKF operations, about 100 AK-47s per Commando Battalion were eventually provisioned that met the requirement of just about one assault team.

The terrain offered a different set of challenges for the Para Commandos. Few of the teams had operated in the thick jungles, barring an odd stint or two in the North-East. For most of them, the terrain was the first enemy, leaving them confused and disoriented and it took months of patrolling to sketch their maps. This is where the initiative of the commandos came handy as they adapted to the environment by quickly and steadily plugging their loopholes.

Ironically, Indian media had been sporadically reporting that the nucleus of the LTTE had been trained by the Special Group under the SFF, which, in turn, worked under the external intelligence agency, R&AW. Whatever the case, the LTTE certainly were good fighters with high motivation. The Para Commandos did not have any language training and it took some crucial months to familiarize themselves with the LTTE tactics and methods. Intelligence was always abysmal and was more guesswork than actionable. The commandos were forced to depend on unreliable partners like other Tamil guerrilla groups who were locked in a battle for supremacy with the LTTE. These Tamil groups had little knowledge of the LTTE, without any guerilla and even arms training and therefore, proved worthless in operations. The military tactics of the Tamil Tigers were innovative, leading to initial casualties for the commandos. For most of the Para Commando Battalions, inexperience in thickly forested area and language proved to be a major challenge. Many of the veteran commandos in the junior leaderships had last seen action in the 1971 war. The youngsters who had joined the units after war had barely seen any action, except for the odd operation during *Operation Blue Star, Operation Meghdoot* or in counter-insurgency in the North East. The only thing that they could fall back on was on their innate ingenuity and the desire to add to the formidable reputation of the units.

By the time these mountain Para Commandos landed at Palaly, elements of another Para Commando Battalion from the deserts, commanded by Lt Col Dalvir Singh, had already been placed in Sri Lanka. The third Para Commando Battalion unit was sent to Colombo with orders to protect the Indian High Commission and its High Commissioner, J N Dixit. Both the desert and mountain commando units remained in and around Palaly refining combat skills mutually for impending operations against the LTTE.

Interestingly, the officers and other ranks of the mountain commandos had been put through some tough training phases by the Commanding Officer, who would insist on a high altitude camp every month, and most of the assault teams would be forced to rotate between glaciers and Pir Panjal mountain ranges, doing night marches, practicing ambushes and understanding how to conduct special operations. The intense training helped the commandos develop an instinct for survival. The credo among the commandos was to inflict maximum casualties on the enemy with minimum casualties to themselves. Any casualty, they knew, would add to the enemy's morale and would therefore add to the LTTE's strength. During their deployment in Sri Lanka, this unit also underwent a change in command. Col Pathak had already commanded the unit for over three years and a new officer was waiting in the wings. They chose a quiet veteran of the unit, Col Hardev Singh Lidder, to lead the battalion. Both the Commanding Officers had a distinctive style and left an indelible mark on the battalion. While Col Pathak's unconventional training methods and insistence on perfection prepared the battalion for Sri Lanka, Col Lidder was credited with bringing in "some method to the madness."

He brought in concepts of conducting strategic surveillance on the enemy and to build up a databank of operational intelligence. He insisted that they study the enemy and his tactics thoroughly before launching into operations. As they observed the LTTE, they began to plot its movements on maps and adding to the battalion's institutional memory. This helped the unit to inflict maximum damage to the enemy with minimum casualties. They would reconnoiter an area before setting up an ambush and use different routes to arrive at the objective in very small teams. After a successful operation they would exercise caution in exfiltrating from the area of operation by taking different routes back to the base. This tactic helped minimise casualties in comparison to their sister unit from the deserts.

The desert commandos had an experienced Commanding Officer who hailed from Rajasthan. Lt Col Dalvir Singh was known as one of the toughest soldiers in the army with years of combat experience. He had been one of the key group of officers who had formed the nucleus of the National Security Guards that had been raised for counter terrorism operations. He was attached to the 52 Special Action Group (SAG) that would take over all anti hijack tasks while its sister battalion, 51 SAG would be responsible for all other direct action tasks against terror attacks. During the successful *Operation Black Thunder-II* when the military intervened in the holy shrine of the Golden Temple, Lt Col Dalvir Singh had insisted on taking part in the operations. Although suffering from asthma, the officer would lead operations, constantly looking for opportunities to send in his assault teams for operations.

The third Para Commando Battalion led by Col Arjun Katoch was posted in Colombo. However, his tenure as the Commanding Officer was coming to an end and a new Commanding Officer was expected to fly in and take over the battalion. His successor would be Lt Col Prakash Katoch, a veteran of the 1971 war and the team commander in urban counter-terrorist operations for which he had been awarded the Shaurya Chakra for gallantry. By the time Lt Col Prakash Katoch assumed command of this "youngest" of the Para Commando battalions, the unit was getting to grips with the LTTE having moved out from Colombo. Its assault teams were deployed over three Divisional Sectors including Batticaloa, coastal town of Trincomalee and heavily forested Vavuniya, the assault team in Vavuniya working closely with the mountain commandos under Lt Col Hardev Singh Lidder. Most of their tasking was in support of the respective Infantry Division in whose sector they were deployed. They would conduct aggressive reconnaissance and surveillance operations and then set up ambushes. A number of search and destroy missions were also launched by these commandos as the battalion moved from one operational area to another, apprehending / killing militants and recovering considerable quantity of weapons. This unit, led by Lt Col Prakash Katoch with an assault team from the mountain commando battalion, also spearheaded the massive 31 day IPKF operation involving several battalions to trap LTTE Supremo Prabhakaran in the dense Alampil Forest. But, while they got hold of the main LTTE camp replete with underground tunnels and shelters, Prabhakaran had melted away since the cordon around the massive forest was built on foot and by helicopters by several infantry units over a period of days that left little room for surprise.

Sri Lanka was also a time to experiment with and a number of unique initiatives were encouraged to be undertaken by the senior military leadership to aid the Para Commando operations. Lt Gen AS Kalkat, Overall Force Commander and GOC IPKF had strongly recommended to Army Headquarters that the Para Commandos be renamed as "Special Forces" and had issued instructions to the Divisional Commanders not to use these special troops as regular infantry especially for frontal and direct assaults. This recommendation of Lt Gen Kalkat was to sensitise the military hierarchy for making a clear distinction between the Special Forces and the regular Parachute Battalions or any other commando type of forces. His recommendation was the very basis for eventual change of nomenclature from Para (Commando) to Special Forces.

For the first time the special operations cell, then known as the "Commando Cell" in Military Operations Directorate at Army Headquarters insisted on moving to Sri Lanka to help coordinate special operations of the three Para (Commando) battalions. The man responsible for this was Col Rustom Nanavatty, Director Commando Cell, who had participated in the three-man committee set up to reorganise the Special Forces. Col Nanavatty had made detailed notes during his study and his earlier visit to the UK as the IALO and the time had come for him to put his unconventional ideas into motion. This would be a time of great change for the Para Commandos as they began to work like special operators and graduate to becoming Special Forces.

When he arrived on the island-nation, Col Nanavatty found that most of the Para Commando units were employed primarily on "Priority-II" roles. This meant that they were performing mostly tactical tasks in support of the formations and units deployed on the ground. This led to a constant clamour among the formations for more and more Para Commandos, stretching them beyond their limits. It also led to a lot of operational chaos, which was not a new phenomenon. In fact, this kind of a hap-hazard "demand and deploy" structure of the Para Commandos would continue to haunt them in future combat situations including in various counter insurgency theatres, specifically in J&K.

Col Nanavatty knew that to prevent injudicious use of the Para Commandos, it was imperative to set up a centralised command with the blessings of the highest Headquarters. This would help in re-orienting the Para Commandos from Priority-II tasks to Priority-I tasks. In turn, this would result in operational cost effectiveness and lethality.

In Sri Lanka the problems of command and control were exacerbated owing to the physical separation between Headquarters Overall Force Commander (OFC) IPKF, the Headquarters Para Commando Task Force and the Para Commando Battalions and the less than satisfactory communications support. So, Col Nanavatty sought to create an *ad hoc* Headquarters Para Commando Task Force to ensure better coordination between the three units. He recognised that the Para Commando operations would remain largely small scale.

Many field commanders viewed the Para Commandos as troops to conduct rapid hit and run offensive actions. This was detrimental to the Para Commandos and led to a neglect of vitally important reconnaissance and surveillance tasks. There was a tendency to use the units to react to every bit of information in the mistaken hope that out of the hundred attempts one might result in success. It was also observed that in theory, the Priority-I tasks for the Para Commandos in *Operation Pawan* could have included:

▸ Covert operations involving the use of civil vehicles and dress as well as using captured weapons to confuse the enemy.

▸ Coordination, direction and control over the groups that were rival to the LTTE.

▸ Conduct pseudo gang operations.

But these tasks were inhibited by several factors:

▸ There was a lack of clear cut direction and policy at the highest level in the conduct of rival groups and pseudo gang operations.

▸ There was a lack of intelligence at all levels and insufficient integration of agencies.

▸ There was a lack of training and skills as well as experience and this resulted in a lack of confidence in conducting such operations.

At the time when the IPKF went into Sri Lanka, the contingency of fighting the LTTE was not visualised either by the government or the military. Eventually when they saw merit in targeting the LTTE's top leadership, this intention was not matched with clarity with regard to the means and methods when it came to the employment of the Para Commandos. To eliminate the top LTTE leadership there was requirement for the conduct of precise, surgical operations which fall in the category of offensive intelligence operations. The absence of an effective intelligence organisation and support both from within the military and outside it, led to a major handicap for the Para Commandos.

Instead, the Priority-II tasks took over and the Para Commandos were used to achieve them. These were:

▸ Search and Destroy operations, distinct from deliberate cordon and search operations. The Para Commando sub units were inserted into pockets of continuing LTTE activity with the aim of searching for contact and destroying the enemy.

▸ Inland Water Interdiction Operations were undertaken to dominate the Jaffna lagoon.

▸ Operations were also conducted in support of the Infantry brigades with mixed results.

▸ 'Snatch' operations at the squad and troop level involving surveillance and selective apprehension of suspects in built up areas by night were also conducted.

▸ The cordon and search of island territories fell to the Para Commando units where they would conduct deliberate screening of the population to identify and isolate LTTE cadres.

▸ Surveillance, ambush and strike meant long range patrols and penetration of the jungle. These came closest to Priority – I tasks along with the 'snatch' operations that were also undertaken by the Para Commandos.

In fact, at the later stages the Para Commandos borrowed from the Rhodesian Sealous Scouts and created two special squads for the conduct of covert operations. However, the lack of coordination between operations of holding formations and the Para Commando units was often in evidence. This led to "blue on blue" clashes since the Para Commando operations were primarily based on stealth and secrecy leaving the super-imposed conventional troops to view them as the "enemy". Therefore it would have been imperative to ensure that the holding formations were not super-imposed on the areas where the Para Commandos were operating. There was also a problem of turf and game of kills, for instance, one Infantry Brigade Commander (ironically himself from the Parachute Regiment) told a Para Commando Battalion Commander that he will not permit the Commandos to operate in his Brigade Sector. Eventually, the latter had to approach the Divisional Commander who promptly overruled the Brigade Commander.

A major problem that the Para Commando units faced was their theatre-specific roles. While one Para Ccommando unit had specialised in the mountains, the other one had been designed to operate in the deserts. This proved to be a major obstacle as a few individuals failed to come to terms with fighting in Sri Lanka. They would constantly hark back to their theatre specific combat roles implying that *Operation Pawan* was an unnecessary and entirely avoidable interlude.

Despite these institutional obstacles, several experiments were conducted with reasonable success. A Special Group was created by taking officers and men from two Para (Commando) Battalions and christened the Tiger Troop. The objective of Tiger Troop was to conduct intelligence based operations and serve as a quick reaction team to neutralise the LTTE leadership. Tiger Troop met with limited success when they captured Jyothi, a top LTTE commander who was also responsible for one of the LTTE radio stations. A message was passed on to the other LTTE commanders to gather for a meeting. This elaborate ruse was being worked out to ensure that the LTTE leaders could be drawn out and neutralised but the ruse fell apart at the last minute. However, Tiger Troop did interact with several Tamil guerilla outfits and undertook special operations based on the specific intelligence that they had gleaned from the leaders of the rival groups.

But the most ambitious action that was undertaken by the Para Commandos was in October 1987 when a team from the desert commandos was tasked to take out the LTTE leadership in Jaffna. Intelligence from R&AW had come in that the top LTTE leadership, along with V Prabhakaran would meet near the Jaffna University in October for a major conference. The commandos were tasked to conduct a heli-borne operation with support from a company of an infantry battalion which would provide security to the helipad as the commandos went in quickly and neutralised or captured the LTTE leadership.

Lt Col Dalvir Singh was in Colombo and he asked his Second-in-Command, Major Sheonan Singh to set up an assault team under Captain Rajiv Nair. Major Sheonan was new to the unit and therefore it was felt that Captain Nair would be the Team Commander and lead the assault. Captain Ranbir Singh Bhaduria would be the Troop Commander for the Special Forces team that would fly into Jaffna in the lead helicopter. Four Mi-8 helicopters from the Indian Air Force were the only available resources to carry in the commandos and the infantry company. This would mean over twenty sorties into LTTE held territory and would result in total lack of surprise with every subsequent sortie after the initial ones. Incidentally the four Mi-8s were from three different helicopter units and had not worked together before. All of them were used for routine tasks and none of the pilots had ever trained, let alone participate in a special operation.

On night October 11/12, 1987, the Special Forces team was tasked to secure a landing zone for the Infantry Battalion, then raid the LTTE political headquarters or any other target in the vicinity and subsequently link up with the Infantry Brigade at Jaffna bus stand by first light or establish a commando base till link up by the Infantry Brigade took place. This was the first operation for the Para Commandos and they knew virtually nothing about the enemy's fighting capabilities, his firepower and his tactics.

Major General Harkirat Singh, the GOC of the Infantry Division in Jaffna and his Col (General Staff) Col Hoshiar Singh gave the initial briefing for the heli-borne operation and rappelling down was ruled out to prevent casualties. Each Mi-8 could carry a stick of twenty troops with their equipment and had to carry a total of 480 troops. A football field without any features that could offer a minimum cover was identified as the landing zone (LZ) for the operation.

At 0100 hours, there was half-moon in the sky and the flying time to the LZ was of barely two minutes. The first few waves would see the Para Commandos being landed in a football field followed by the company of infantry troops. While the Mi-8s managed to land the majority of the Para Commandos, it was realised at the last minute that the none of the infantry troops were ever trained for heli-borne operations. They had just come in from Gwalior and worked in the conventional format, carrying huge loads of arms, ammunition and equipment that was useless in an operation of this nature. It meant reducing the number of troops which could be carried in each sortie and led to confusion in the heat of battle. Before the first two sorties had been completed the LTTE cadres had got wind of the assault. They immediately rushed reinforcements and placed their MMGs and rocket launchers on high rise buildings and began to target the Mi-8s as soon as they came within range.

It took the Para Commandos just a few minutes to be over the Landing Zone (LZ). Capt Bhaduria was at the door with the rope in hand as the helicopter lowered down to 10 metres. He threw the rope down the rope and made a quick exist, holding the rope tight for the next man to come down. The LTTE immediately opened fire from the medical university second floor dominating the landing ground. The pilot decided to land the helicopter instead of hovering it at 10 metres, making it an elusive target for the enemy. However due to the down draft of the rotors the lengthy rope got entangled around Captain Bhaduria's legs. As the helicopter landed, he could see the last commando rushing out of the helicopter with the second in command bring up the last man. Captain Bhaduria had two choices. He could either jump back into the helicopter and come back with the next wave, or else go back hanging under the helicopter and then come back with the next wave.

His commando dagger was in his rucksack which could not be taken out unless it was released. But going back into the helicopter was the last option and going back hanging under the helicopter's undercarriage was a difficult option with a 25 kilo back-pack. But luck was on his side and he managed to untangle himself just as the last commando was moving out of the helicopter.

Some of the Mi-8s were forced to turn back and lack of communication between the pilots from the different helicopter units led to further confusion on ground, stranding the Para Commandos. But their superior training helped them fight their way into the surrounding buildings in the university and hold out by conserving their ammunition. However, troops of the infantry unit who had landed at the football ground to maintain the security of the LZ came under withering fire from the LTTE and with no cover they were cut to pieces.

It was the third wave of two MI-8 helicopters that brought in Major Virendra Singh from the infantry unit with his Mortar Platoon and Medium Machine Gun (MMG) section. But in the heat of battle the infantry platoon had left the mortar ammunition behind and the whole platoon was rendered useless without their primary support weapon. Only one of the men out of the 29 survived and he too was captured.

When Lt Col Dalvir Singh heard of the disaster he took the first flight out of Colombo and landed in Palaly and with a small team of commandos set off to rescue his men from Jaffna. Along the way he spotted three T-72 tanks led by Major Anil Kaul from an Armoured Regiment. Lt Col Dalvir Singh studied the maps and realised that a railway track ran parallel to the University where his men had holed up and were surrounded by the LTTE. He chose to fight his way into Jaffna using the railway tracks as his infiltration route, supported by the three tanks. Unfortunately, LTTE cadres managed to target the lead tank carrying Maj Anil Kaul and fired several rockets at him that incapacitated him as he lost one eye. Lt Col Dalvir Singh, who had never operated tanks had to take over command and direct them as they continued to fight their way into Jaffna. After two days of heavy fighting since the operation was launched Lt Col Dalvir Singh managed to link up with his beleaguered assault team and bring them back to Palaly.

This operation would be one of the biggest failures for the Para

Commandos and one of the most tragic episodes in the history of the Indian Army. Once again the failure of the senior military leadership to appreciate the intricacies of high risk special operations had led to this tragedy. Intelligence agencies had provided sketchy and wrong inputs which included unsubstantiated information that the LTTE cadres in Jaffna were "lightly armed." These proved to be critical failures that were coupled with the unfamiliarity of the helicopter pilots with the needs of Special Forces. It was left to their courage to complete the mission and the helicopter pilots displayed tremendous guts as they took in the Para Commandos under withering fire from the enemy to ensure that they complete their mission. Perhaps the failure at Jaffna University facilitated the arrival of Col Nanavatty a month later starting a fresh round of focused operations and an attempt to break out of the Para Commandos mould and begin the long road to becoming Special Forces.

Sri Lanka would become a turning point in the evolution of the Indian Special Forces. They come to face with a competent, hostile and ferocious enemy and were forced to adapt to their tactics and left an indelible mark on the LTTE. As per Maj Gen Shiv K Sharma, then IG SFF, LTTE radio intercepts clearly indicated that they feared the Para Commandos most and chose extra caution whenever the slightest movement of these specialist troops was apprehended / detected in any area. A number of joint operations would also be conducted by Army Para (Commandos) and the fledgling Indian Marine Special Forces (IMSF) after two naval officers returned from the BDS/SEALS course in the United States to start a project for the Navy's Special Forces.

While the Para Commandos were engaged in battling the LTTE, Brig Michael Rose, Director, SAS from UK came and met Lt Gen AS Kalkat, OFC & GOC Land Forces. Rose subsequently visited Sri Lanka to gain firsthand knowledge of the ongoing special operations. He also invited Gen Kalkat to visit SAS, which the General did at a later date, discovering amongst other things that the SAS had no formal rank structure. Brig (later promoted to General rank) Michael Rose was the Commanding Officer of the SAS when they went into the Iranian Embassy in London when it was taken hostage by terrorists. The hostage rescue operation would become a classic text-book case on how to conduct such special operations and bring the SAS into the limelight for the first time. The world would wake up to the existence of the shadowy SAS as the men conducted a multiple entry rescue operation, neutralised the terrorists and captured one of them alive. He would subsequently go to command British Forces in the Falklands and also in Bosnia.

The experience gathered in Sri Lanka would leave a major mark on the Para Commandos and by 1990, when the last elements were preparing to return to India, they had learnt several valuable lessons. Officers would joke about the key differences in the way the three battalions would operate. The mountain commandos would have the least casualties as they would operate in stealth. They would take different routes to an ambush, spend considerable time conducting intelligence surveillance and then carry out their ambushes. They would then take a different route to extract themselves after the action was over so that they could not be ambushed as they returned to base. In sharp contrast, the desert commandos would go into operations with courage and bravado that would yield results but also lead to several casualties. Much of this would end up becoming part of the military yore that would dominate the Sri Lankan narrative in the evolution of the Indian Special Forces.

Many of these lessons would come handy as the units were drawn into ferocious counter insurgency operations within months of their return. A botched election in the State of Jammu and Kashmir had erupted in wide spread protests and Pakistan was waiting to exploit such a situation. Pakistan, as a state had always believed in low intensity conflict as a means of its foreign policy and it was all set to begin a new war with India. Once again, the Para Commandos, now re-organised as Special Forces would play a leading role in taking on cross border terrorism; fighting terrorists who had forged their expertise in battling the Soviets in Afghanistan.

A new era and a new challenge for the Indian Special Forces was all set to begin.

Section II
The Special Forces Years
(1990 and Beyond)

1

Op 'Rakshak' - Counter Insurgency Employment

By the time the three Special Forces units were coming back from Sri Lanka, a major insurgency was erupting in Jammu and Kashmir. The state administration was caught unawares as militants from the Jammu and Kashmir Liberation Front (JKLF) began to infiltrate across the LC after receiving training in Pakistan. These were early days and the JKLF was the dominant insurgent group in the Kashmir valley, made up of indigenous Kashmiris who had crossed over to Pakistan a year before. The subsequent years would see a marked change in the insurgency as the complexion of terrorism changed in the Valley from a purely indigenous movement to one that would soon be dominated by foreign militants who were moving out from Afghanistan and re-routed to India by Pakistan.

This was a legacy of the Soviet invasion of Afghanistan in 1979, when the tanks rolled across the Hindu Kush and set up a communist regime in Kabul. The CIA immediately forged a relationship with the Pakistan's Inter-Services Intelligence (ISI) to start creating a *Mujahideen* that would take on the Soviets. The money for the insurgency would be funneled through the ISI, which would choose the groups they wanted to fund or arm. The *Pashtun* groups would receive the bulk of the money while the minority rebel groups of the Tajiks and the Uzbeks in the North would get a lesser dole. By the late 1980s, Indian intelligence surmised that the infrastructure that had been created by the ISI for Afghanistan had become useful to Pakistan for starting an insurgency in Jammu and Kashmir. The plan was not new and had been attempted in 1947, soon after Partition of the

Subcontinent, when Pakistani Army regulars dressed as tribal raiders took off from Domel and launched a major assault on Kashmir. After the accession of Jammu & Kashmir to India, Indian troops landed at Srinagar airport to save Kashmir.

But in 1990, the same strategy had become far more potent, funded by the CIA's funds that were being diverted to fuel a new insurgency. The state administration watched helplessly as the militants ran amok across the state even declaring some regions as "liberated." It took the security forces nearly four years to regain a foothold and begin to take on the militants. By 1994, when Gen B C Joshi took over as the Chief of Army Staff, the contours of a fresh strategy to deal with the terrorism in the Valley began. Gen Joshi, a visionary Army Chief who had served as DGMO, facilitated a comprehensive study to reorganise the Special Forces and begin a process to modernise the Army.

It was his plan to create a permanent force that would be manned by officers and men from the Indian Army on rotation but permanently stay in Jammu and Kashmir to tackle the growing insurgency. Naturally, the mountain commandos were to play a major role in the counter insurgency operations.

This was also a time of great churning in the three units that had just returned from Sri Lanka. They had been blooded in one of the most violent insurgency campaigns ever conducted and had met with mixed success. But the biggest gain for the units was the ability to prepare themselves for new conflicts and adapt to various situations by breaking out of their theatre-specific roles. They were no longer Para Commandos and Sri Lanka had provided them the opportunity to ensure that they could take on any enemy, on any turf and deliver a creditable account of themselves.

There was also a generational shift in the three battalions as new officers, who had joined the unit after *Operation Pawan* were getting oriented towards a new low intensity conflict. The lessons that their seniors had learnt on the battlefields of Jaffna, Trincomalee, Batticaloa, Palaly and Vavuniya would help in the new battles fought in Sopore, Lolab Valley, Doda, Kishtwar, the Harfuda forests and Kupwara. The tactics were the same, as the Special Forces began to adapt to the new enemy, learning their customs, language skills and matching their weapons with some of their own.

A marked change in Sri Lanka was the authorisation to start using the captured AK-47s from the LTTE. The units had come back with a sizable number of the relatively modern and sturdy assault rifle and orders were placed to get more AK-47s from the Warsaw Pact countries. But a price-sensitive Ministry of Defence looked for the cheapest rifles in the market and when the first MK-VZ version of the rifles arrived, the units found that their rifling had been worn out with usage. The bullets would fire of in different directions and they had to wait till better versions of the weapons arrived. But the coming of Gen B C Joshi proved to be major advantage for the Special Forces. A new regular battalion was being earmarked for conversion into a Special Forces role and there was movement to set up a Special Forces HQ along with a Special Forces Regiment.

A new generation of young officers had come into the battalions, some of them who had joined their battalions in the final eight months of their deployment in Sri Lanka. Officers like Lieutenants Abhay Narayan Sapru, Dalip Bhalla, G S Kullar, Arun Singh Jasrotia, Alok Jha and Lt Sudhir Kumar Walia made it through the rigorous probation of Special Forces. Lt Sudhir Kumar and Lt Jasrotia would become legends within the Indian Special Forces community in a few years, as they took on the Pakistan sponsored terrorists in the Valley in continuous operations.

Both Lt Walia and Lt Jasrotia joined the Special Forces almost together, separated by a few months. Lt Walia was commissioned as a part of the 82nd Regular course and six months later Lt Jasrotia would also receive his commission as a part of the 83rd Regular course. Both would be part of the new breed of young officers who would bear the brunt of the new wave terrorism engineered by Pakistan that was spreading across the Valley.

Captain Alok Jha, a course junior to Captain Jasrotia joined the Special Forces unit in the mountains a year before him. Captain Jha would conduct the probation for his "senior" Captain Jasrotia when he would show up as a volunteer for the unit. From the very beginning there was something unique and different of this quiet man who would go about the tasks set to him with ease and elegance. Part of the probation was a 60 kilometres speed march with 40 kilograms of weight that had to be completed in a set time frame to test the stamina and determination of the volunteer aspiring to join the Special Forces. Nearly eight kilomteres into the speed march, the strap on of the back-pack of Captain Jasrotia's would snap off. Captain Jha, who was the Probation Officer offered him a chance to conduct the speed march at a later date so that the back-pack could be repaired. But Captain Jasrotia just refused to stop and finished the speed march well in time carrying his bulky rifle in one hand and the back-pack in the other.

He and Captain Jha would become best buddies as their unit got sucked into counter insurgency operations. The operations carried out in those years were unique and presented fresh challenges for the Special Forces. In those days, parts of Kashmir were considered "liberated" by the terrorists flooding in from across the Line of Control (LC). Sopore a small town in the Valley was one such "liberated" town (as the army had no permanent presence, after each operation army moved out, which permitted the militants to re-enter) when their team was tasked to take on the militants. Captain Jasrotia would earn his first Sena Medal for gallantry and they went into the operation fighting their way into built up areas against known militant haunts. A particular area was notorious where the militants would simply fire at a military convoy and melt into the crowd making it impossible to fire back. A plan was quickly made to send in two teams, one covertly and dressed in civilian clothing while another team would drive in as "bait" for the militants. Major Paramjit Sangha, the Team Commander immediately decided to act as the "bait" and drive in with some of his men hoping that the militants would squeeze off a few shots.

At first light, as the Special Forces team merged into the town, Major Sangha drove and began to make the rounds hoping that a militant *tanzeem* would take a shot at them. As expected, as soon as a militant team opened up, Captain Jha and Jasrotia's troops moved in and gave chase. The militants ran into a nearby *dhok* (Kashmiri hut) and hid in the false ceiling that is customary in these houses.

Captain Jha ran into the *dhok* quickly followed by Captain Jasrotia and another Special Forces operator when the militants fired a short burst and threw an *Arges* grenade, the standard munitions that militants came equipped with from across the LC. The grenade bounced off Captain Jha's chest and fell between him and the entrance where Captain Jasrotia was standing. With nowhere else to go, Captain Jha would jump inside with his back to the grenade to avoid any major injury while Captain Jasrotia rolled away from the door. Luckily for both officers that grenade would be that rare piece that wouldn't go off. In that operation the team would get nearly six militants, a rare find in those early days of insurgency in the Kashmir Valley.

Promoted to a Captain, Jasrotia was chasing a group of terrorists on September 15, 1995 when his team came under withering fire in the Lolab valley. Capt Jasrotia had already been awarded the Sena Medal for gallantry and along with Lt Sudhir Walia, both considered the new breed of Special Forces officers in the Indian Army. That day Capt Jasrotia's team came under heavy rocket and small arms fire. In order to extricate his men, he crawled forward and despite suffering severe injuries, he killed a terrorist with his commando knife in hand-to-hand combat. He managed to kill another terrorist by lobbing grenades. This gave his team an opportunity to take cover and engage the remaining terrorists effectively. In the ensuing firefight Capt Jasrotia succumbed to his injuries. For his gallant action Captain Jasrotia was awarded with the highest peace time gallantry award of Ashok Chakra posthumously. Captain Jha was away in Nepal during this time. A call from the unit to Major Sangha's relatives in Kathmandu would break the news to him. It would take a few days for him to book a flight back to Delhi, landing on the day when the final remains of Captain Jasrotia were consigned to the flames.

A few years later, on August, 29,1999 just a few months after the Kargil war, Major Sudhir Kumar Walia was on an intelligence-based operation with a squad of five SF personnel in the Hafruda Forest in Kupwara district. The squad suddenly came across a well camouflaged hide out housing nearly 20 terrorists. Major Sudhir Kumar pushed ahead of his squad and taking advantage of the element of surprise, neutralised the sentries and killed four militants. However, he had also received serious injuries he continued to engage the terrorists. Having lost a lot of blood, Major Walia continued to direct his men on his radio set and ensured that the terrorists were eliminated. Like his buddy Capt Jasrotia, he was also awarded the highest peace time gallantry award of the Ashok Chakra posthumously.

In March 2007, came another gallant act and sacrifice by Captain Harsharan R in close quarter combat with terrorists in north Kashmir. On receiving information in the dead of night of presence of hardcore terrorists in a house he laid a cordon. In the early hours of the morning, four terrorist in the curtain of dark and heavy snowfall rushed out firing simultaneously, virtually face to face with Harsharan. Despite being outnumbered, Harsharan killed one terrorist instantly but received a bullet in his thigh. Unmindful, he killed a second terrorist rushing him but received a gunshot wound in the neck but wounded a third terrorist. Harsharan later succumbed to his injuries. He was awarded the highest peace time gallantry award of the Ashok Chakra posthumously.

In yet another audacious counter terrorist operation in 2009, after receiving information of presence of some infiltrating terrorists, Major Mohit Sharma (already a recipient of gallantry award Sena Medal) led his commandos in tracking them in the dense Haphruda forest of North Kashmir. On observing suspicious movement, he alerted his scouts but terrorists fired from three directions indiscriminately. In the heavy exchange of fire, four commandos were wounded immediately. With complete disregard to his safety, he crawled and recovered two soldiers to safety. Unmindful of the overwhelming fire, he threw grenades and killed two terrorists but was shot in the chest. In the brief respite that followed, he kept directing his commandos despite serious injuries. Sensing further danger to his comrades, he charged in a daring close quarter combat killing the balance two more terrorists and attained martyrdom fighting for his motherland in the finest traditions of Indian Army. He too was also awarded the highest peace time gallantry award of the Ashok Chakra posthumously.

Insurgency settings provided the Special Forces fair opportunity to operate in high mountain ranges, dense forests, villages, as also thickly populated areas. Invariably in latter cases, the call came when terrorists had already taken hostages or opened fire from a safe house on their targets in a surprise move or on being discovered. Such occasions were many and gave youngsters like Majors Manav Yadav and Puneet Atwal opportunity to display their skills, eliminate the terrorists and free hostages. Recognition and awards kept pouring in routine fashion.

While the Rashtriya Rifles (RR) was being set up, Gen B C Joshi was also keen to set up a commando RR unit that could emulate the Special Forces and conduct surgical operations against the terrorists. Colonel Ram Pradhan, a veteran Special Forces officer was brought in along with Major Abhay Sapru, a veteran of the Sri Lanka operations. Most of the men were either from the Dogra Regiment or the Jammu & Kashmir Light Infantry. This proved to be an interesting challenge for the Special Forces officers who had been brought in hastily to try and shape this fledgling unit into a battalion fit for operations. Most of tactical training given to the troops was drawn on the experience that Maj Sapru had picked up as he had operated in Sri Lanka.

The unit had to start from scratch, desperately trying to achieve cohesion in a limited time frame as most of the men and officers would be sent on deputation to the RR and then rotated back to their parent units. The men had to un-learn their conventional tactics, learn new skills akin to commandos and operate using unconventional methods. This is where the experiences gained from Sri Lanka came in handy and the troops began to shape up and conduct their operations, starting with the Doda valley in the south. The unit was involved in a major operation during the period when a new militant outfit, Al-Faran, kidnapped several foreign tourists in the Kashmir Valley. While the Army and the RR Battalions were immediately pressed into service, a team from this unit operating in Doda, south of the Pir Panjals in Jammu district received information that a militant identified as "Sohaid Bhai" was looking for contacts in the Kishtwar area. There was another terrorists with him – Hasan Ali also known as Al Turki. They had been tasked by the Al Faran to establish contact with the Harka-ul-Mujahideen squad in Kishtwar so that the hostages could be moved there to escape the army units searching for them. The unit established a dragnet across the area preventing the two militants from establishing contact as a cat-and-mouse game continued over the next 72 hours. Worried that commandos would get them, a Gujjar guide who was working with the militants ran away with Hasan Ali's weapon. The commando team finally chanced upon the two militants who immediately gave up and led them to their hideout where two cameras of the hostages were recovered, details and whereabouts of the hostages who were still in the Kashmir Valley, however, could not be ascertained.

By 1991, teams from other Special Forces units outside J&K were also being inducted into the Kashmir Valley. Capitalizing on surrendered militants, Special Forces teams were creating small irregular bodies spear headed by Special Forces operatives. These irregular bodies of former militants were created without any major training, equipment or tasking. This training for pseudo-gang operations was as had been envisaged by the Para Commandos while dealing with the LTTE in Sri Lanka.

They worked on the fault lines between the rival militant groups and began to look after the families of those who had signed up for the effort. This paid dividends as the first major special operations against the terrorists swung into action. The troops stayed with them in their homes, did ambushes together and began to live like them. These surrendered militants would provide key information on terrorist outfits like the Al Badr and help the Special Forces patrols neutralise several prominent militant leaders. They also helped prepare a major disinformation campaign that was created by the Special Forces as they operated in the interiors of the State. In fact post 1997, whole Special Forces units from outside J&K started moving in for a few months each by and operated effectively, killing scores of terrorists. Old veterans of the Sri Lanka operations like Colonels Mohan and Sanjay Bhaduria would use the advanced LTTE radio sets captured during *Operation Pawan* to scan the airwaves for militant communications and ambush them.

It is not that these Special Forces units had to depend on surrendered militants to undertake pseudo gang operations. Over a period of time most Special Forces operatives could very well merge with terrorists, speaking their language, dressing and living like them. It became a fun game for youngsters like Capt KP Singh to meet infiltrating Pakistani terrorists and scalp them. His unit conducted several special operations as they quickly learnt the language and customs of the foreign militants. They would spend time posing as a militant group and establish contact with other militant groups to gather intelligence. Once they had gathered adequate intelligence they would either ask the militants to surrender or engage them in firefights, when met with resistance. Most of these skills took a lot of time and patience and brought to the fore how effective Special Forces could be in counter-terrorism operations.

The same unit would also undertake several successful counter-insurgency operations in the North East. An operation conducted by a small team of two young officers and two NCOs in a covert mode in March 1998, in general area Churachandpur of Manipur in Counter Insurgency environment proves this. The initial plan was made when an input was received that a local resident in a remote village close to Myanmar border produces country made AK-47 weapons for the militants. Based on this information, this small team planned a snatch operation to pick up this individual to carry out interrogation to gain more useful inputs. This small team moved in covert mode in a civil hired van to the intended objective, however, due to heavy rains and poor road conditions, the operation was called off midway. The team leader decided to take a different route for coming back to base, which although short, was not used by military convoys and accordingly, there was never a route opening done for this route. The team had just traversed a few miles on this stretch, when they sighted a column of 10-12 militants walking in a formation with weapons and combat loads, confidently, as Army carries out patrolling. Movement of such a large group of militants was probably due to army presence being minimal in that area. As the van crossed this column, militants tried to stop this van, probably for checking or for extortion. As the team leader did not stop the vehicle, few shots were fired on the van, which only damaged the rear windshield.

The team leader halted the vehicle after going about a kilometre ahead and made a plan to apprehend this group in spite of being totally outnumbered in manpower as well as weapons, as such a lucrative target is a rare opportunity to come by. The team climbed a hillock after hiding the van in thick undergrowth. The opportunity ambush was so sited to cover open flanks also, as it was envisaged that on contact militant might try to manoeuvre around the ambush site. Armed with just two AK 47 with five magazines each and two pistols each round had to make hit the militants. As the column of militants reached in close vicinity, they were effectively engaged by the team. The firefight lasted for about 20 minutes after which anticipating a strong resistance, the militants retreated. The team waited for darkness to undertake the search of the area. The success of this operation included the neutralising of five militants with weapons and a large quantity of explosives and incriminating documents. This was a classic small team operation by a Special Forces team highlighting initiative at junior leader level and a bold planning and execution.

After the Kargil Conflict Special Forces units from outside J&K continued to operate in the Valley, notching up significant successes. In one instance, intelligence had come in from intercepts that a senior militant commander of the Lashkar-e-Toiba was moving around in high mountains area north of Srinagar. A Special Forces team was immediately airlifted and placed in the area near the last sighting of the terrorists. The unit was known to improvise their standard issue equipment and they had hooked up their hand held thermal imagers to locally procured TV sets to monitor the house where the terrorists were holed up. A few days later, as the militants moved out, a small Special Forces team slipped out and began to trail them until a firefight broke out in a nearby *nala*. The firefight resulted in both the terrorists being neutralised. One of the dead terrorists was identified as Kasib, the son of the operations commander of the LeT, Zaki-ur-Rahman 'Lakhvi'. It was Lakhvi who would later plan the assault on Mumbai on 26/11, 2008.

By 2003-04, the Special Forces would play a leading role in the counter terrorism operations in J&K state. Veteran Special Forces officers like Hardev Singh Lidder, who had now taken over a RR Force as a Major General would oversee one of the most successful operations against the Pakistan sponsored terrorists in the upper reaches of Hill Kaka, a treacherous mountain region to the west of the Poonch-Surankot Sector. The area had become a permanent base for the terrorist infiltrating across the LC from Pakistan Occupied Kashmir and gather at Hill Kaka before moving into the Kashmir Valley or Doda / Kishtwar. An elaborate operation was planned for over a year with a Special Forces Battalion playing a leading role. They also picked up a number of Gujjar boys who were familiar with the terrain and trained them. In May 2004, one Special Forces team launched the first assault on Hill Kaka and destroyed the base that had facilitated the infiltration of hundreds of militants for over a decade.

The involvement of the Special Forces in counter insurgency operations in Jammu & Kashmir proved to be an excellent training ground but also came at a considerable cost to their capabilities. All the three Special Forces Battalions had also repeatedly demonstrated their level of skills at the individual and unit levels. The officers and men had continued to expand their skills at an impressive rate with many of them specialising a vast array of skills that included conducting High Altitude High Opening (HAHO) and High Altitude Low Opening (HALO) Combat Free Falls (CFF), deep sea diving and other specialist capabilities. Unfortunately, in the higher echelons of the military leadership, there was little or ill-informed awareness about these skills or their capabilities. As the Special Forces continued to grow, they would be either misused or be subjected to impossible expectations, leading to continuous clashes that would be to the detriment of the Indian Special Forces.

While the military recognised that the Special Forces had a major role to play in modern warfare, amply demonstrated by the US invasion of Iraq in 1990 and again in the aftermath of 9/11 in Afghanistan, there was little movement in the direction that would have ensured the Special Forces achieve their true potential.

The unfamiliarity of the military hierarchy with special operations would show up the fault-lines in operations when war broke out. The Special Forces units, thrown into counter-insurgency operations for nearly a decade also led to a Low Intensity Counter-Insurgency Operations (LICO) mindset in the battalions. By 1999, when war broke out in Kargil, the four battalions were still in a counter insurgency mode. Doctrines had not been updated, internal best practices within the battalions had also eroded with the constant exposure to counter insurgency operations and the military hierarchy had forgotten that the Special Forces were a strategic force, not mere tactical commando battalions.

Part of this was also the internal machinations and a lack of vision that had killed a major initiative that had been started in 1994 by Army Headquarters under Gen B C Joshi. This was the birth of a separate Special Forces Regiment that could have achieved a major milestone in the evolution of the Special Forces and given India a true strategic capability. Instead, what started as a visionary initiative quickly descended into a farce and ended in tragedy.

2

Op Vijay - The Kargil Conflict

The National Highway 1A that courses from the Kashmir Valley and heads north-east towards the mountains of Leh in Ladakh passes through Dras, a small sleepy town that is barely noticed by travelers and beyond to Kargil before entering Leh. From Leh, a road runs North via Khardung La to Siachen region and a separate one to Eastern Ladakh.

Few realised the strategic importance of Kargil and the surrounding heights, most of which are designated as points and numbers where the army undertook patrols or held some features. In winter, because of the sub zero temperatures with some areas comparable with the Siachen Glacier area, troops would vacate their posts and come down to a relatively warmer region. As in all cases of winter vacated posts, there is no question of even an informal understanding between the opposing troops that these areas could not be occupied by the adversary to gain an upper hand. In order to monitor the area, the army would fly helicopter sorties over the area periodically, known as Wide Area Surveillance Operations (WASO) to detect any intrusions. However, these sorties generally flew at a distance of about one kilometer from the Line of Control.

But none of these WASO sorties actually detected what was going on before the onset of summer of 1999; an elaborate plan involving Pakistani Northern Light Infantry (NLI) Battalions and Special Services Group (SSG) sneaking in to capture the vacated posts, plus establish posts of their own.

The purpose of the massive secret Pakistani intrusions was to ensure that National Highway 1A could be effectively dominated and depending on success achieved, cut it off permanently thus gaining a major strategic advantage over India. The Pakistani plan was actually to not lose surprise till the winter of 1999 set in, ensuring that the passes (including Zoji La and Rohtang Pass) along both roads leading to Ladakh were closed for winter. This would have left India with only limited capacity of inducting reinforcement by air during winters. However, an Indian Army patrol was ambushed much earlier that gave away the surprise. By May 1999, when the first intelligence reports from the Kargil Brigade began to filter in, it was obvious that the Pakistanis had established a major foothold in Kargil region. India and Pakistan would go to war for the fourth time and engage in a sharp and bitter conflict that would rage for nearly three months on these icy dominating heights before daring attacks by Indians under withering fire shook the Pakistanis out of the bunkers and sent them packing.

While the bulk of the fighting was fought by the Srinagar based Corps under the aegis of the Northern Army Command, Army HQ was playing a major role. But most of the planning of the immediate battles was left to the Divisional Commanders while assault teams from the various Special Forces units were sprinkled all over the theatre. Once again, the strategic value of Special Forces was not appreciated, leaving them to be used for tactical gains as when a Divisional commander felt the need. The bulk of the fighting was left to the mountain specialists who had always been in the state of Jammu and Kashmir with the other Special Forces units also roped in. But operational matters for the Special Forces would continue to be incredibly complicated. The strategic part in any case became redundant when the then Prime Minister Atal Bihari Vajpayee's Government ordered that Line of Control, as designated during the 1949 Karachi Agreement and 1972 Shimla Agreement, was not to be crossed.

The initial planning as on June 7, 1999 was that a Special Forces team would cross the LC and hold the dominating height Point 5118 on Zulu Ridge. The plan was to land five Troops (each nearly 20 men), and three of them would rappel down on North face of Trishul post and move rapidly along Dukas *Nala* east of Zulu Ridge. The other two troops would cross the Tri-Junction and move to the north along the ravine west of Zulu Ridge. Both these sub units had been tasked to link up by first light, next day at Zulu Ridge from east and west. The main task was to capture and hold Pt 5118. Two other troops were earmarked as reserve for firm bases on Trishul

Ridge to engage the enemy on Yankee Ridge and on Tri Junction.

But the plan changed with the orders having come through on June 16 that all trans-LC operations were to be stood down.

Thereafter, another major Special Forces task was planned and by the middle of the month Operation Brave Heart 1 began to take shape, with the aim to destroy the Pakistani Army's mother base in Gultari region, intruded and established three kilometers inside Indian Territory.

A Special Forces assault team was to move from Sando *Nala* and secure a feature called Sando Top and then move west to occupy dominating height East of Point 4388 by first light on the second day and destroy the mother base by directing Divisional Artillery. Another team was to stay put on Yankee Ridge and be relieved by a company of soldiers from an infantry battalion. The garrison type camp of the enemy was approximately 12 kilometres north-west of Gultari *Nala*. Indian artillery guns were engaging Zulu Top and Tri Junction. But the area known as Black Pimple could not be engaged as the distance was only 300 metres from Indian locations on the Saddle and the firefight was too close to Indian positions in Four Rocks area.

The Pakistanis began to reinforce Sando with about 30 troops by helicopters landing North of Sando. These were believed to be the Pakistani Special Forces, the SSG who took over these positions. The Indians troops had suffered considerable casualties in this area.

On July, 10 the Divisional Commander gave the Special Forces unit a task to terminate the enemy line of maintenance from Gultari Mother Base to Tiger Hill and Point 4875. The approach would be from Sando *Nala* to gain foothold on ridge on Sando South, U Cut, Point 4965. A company of infantry soldiers was to follow and capture Point 5060. The Special Forces unit was to assist capture.

By July 12, when Pakistani forces were teetering, fresh orders were received for the Special Forces unit to capture Sando and the regular infantry were to capture Zulu One Top. By July 15, a plan to capture dominating heights in the area was evolved and the Special Forces unit was brought into *Operation Brave Heart 2*. The aim was to capture Pakistani positions on the dominating heights of Y Ridge. The Special Forces team was to advance along Safed *Nala* northwards and capture a dominating area on Y Ridge earliest. Charlie Team of the Special Forces unit was tasked to carry out the operation and they encountered many Chinese claymore mines along the way. The route to the area identified as Neck and Manali Slope had a three metre wide track that showed the preparations that the enemy had made. Stones had been cut, explosives used and the two kilometers serpentine track was paved all the way, which must have taken a full season (June to August) by a large labour force. Charlie Team moved in and secured Point 5060 and once secured, then began the difficult trek to the area known as 'Neck'. Enroute a Stinger missile was fired at them from a Pakistani *Sangar* (Bunker). On July 17, Charlie Team crossed 'Neck' and got their first view of the Mother Base Gulatri. They saw nearly 20 Pakistani troops approximately 1.5 kilometres in the Mother Base bowl. Two Pakistani helicopters were also seen flying in the area even as Musharraf kept claiming the infiltrators were civilian 'freedom fighters' and not Pakistani Army.

The next day, the Special Forces team occupied the dominating heights of Yankee Ridge, Tri Junction and Sando North. The Gultari gun positions were 14 km to North. Moving at last light along Safed *Nala* they firmed in at Point 5250, by now well within Pakistani-intruded territory; behind enemy lines, so to say. On July 19, the Special Forces Team saw Pakistanis reinforcing and maintaining Point 5070 on Yankee Ridge, Zulu One Top and Tri Junction. Support from Indian artillery fire was not permitted but the Special Forces team was told that supporting artillery fire would be available on call the next day.

By July 20, the weather had packed up with visibility down to zero. The Special Forces Team continued to wait and bunked in for the night. By first light, the Tri Junction Base camp was assaulted and they inflicted heavy damage on the Pakistani position. The Special Forces team was then told to extend the operation to Yankee Ridge for another 48 hours. By July 22, Indian artillery was in action on Zulu Top and Tri Junction. The next day again saw a total whiteout with zero visibility. The Pakistanis retaliated with heavy artillery shelling from the Gultari gun position.

Meanwhile on July 13, the Special Forces unit's Alpha and Bravo Teams were tasked to capture Sando Top and an infantry battalion unit was to go for Zulu One Top. By next day, Alpha Team had established communication relay stations established at Sando Top and Tiger Hill Southeast Spur. Between the intervening night of July 15/16, Bravo Team moved to Trishul Ridge. The firm base was established by last light of July 16 with some enemy positions including Zulu One dominating it. Alpha Team selected and reconnoitered positions for covert surveillance in Sando Valley to suppress enemy at Sando Top. By July 17, regular infantry had moved by day through Sando Valley and having come under intense enemy fire, decided to stay put and sent out a reconnaissance party to Point 4965 and Tiger Hill western Spur. The party reported 18 bunkers from Sando Four Rock area to Zulu Base.

A final effort was made to end this peacefully and the Indians sent across a message for a flag meeting but the flag meeting failed and artillery fire resumed. By this time, Indians were monitoring the Pakistani radio net. On July 20, Indian artillery guns opened up on Sando Top, Black Pimple and Tri Junction. Later at night with Sando Valley approach already well covered by the enemy a plan was prepared for Alpha Team to take the route via Safed *Nala* – U Cut – Sando South approach. Bravo Team would move to firm base to area south to cover Sando feature. By the night of July 21, the regular infantry captured Tri Junction without a fight, Pakistanis having abandoned their position. The next day the Special Forces unit was tasked to capture Zulu One which they did after a fierce firefight the next day.

Most of the Special Forces operations during the Kargil Conflict were similar. They can't be categorised as "Strategic Special Operations" as is understood in the modern context of war. They were good tactical operations, critical to the war effort in a limited theatre, to help Divisional Commanders achieve their aim. This is not how the Special Forces should have been used. But in the absence of a centralised command and control structure, with no strategic aims having been set out by the politico-military leadership, it was left to the Battalion Commanders to say what they could offer, and the Divisional Commanders to rule over what they wanted.

In the nearby Batalik Sector, another Special Forces unit, the Desert Scorpions, had been hastily rushed in from where they were deployed in counter insurgency operations.

In this Sector, one Special Forces team was held back while the Commanding Officer and the other teams were deployed. One assault team of this Special Forces Battalion was used like a company of conventional infantry troops for an assault. The fact that a Special Forces Assault Team is smaller compared to a conventional infantry company and their firepower is meant for special operations, not for full frontal assaults was totally ignored. With lack of basic understanding of Special Forces operations, the Special Forces team was sent into an assault with results that were far from satisfactory. This was true for the other Special Forces teams too that had also been deployed as proverbial fire brigades, set to put out small fires across the theatre of war while the larger picture was lost to everyone. Unlike the US, where a Special Operations Command (SOCOM) or in the UK where the Director, Special Forces would have been in complete control of the Special Forces elements in consultation with the formation commanders, their Indian Special Forces counterparts were left to fend for themselves.

On balance, the leadership in Special Forces was not found wanting. The Unit's *izzat* (Honour) reigned supreme. The physical fitness of the officers, the NCOs and the men was top class. They carried extraordinary loads – self-contained loads for seven days special mission in high and very high altitudes. Reconnaissance and surveillance required over prolonged periods were well undertaken though in on few occasions Pakistani locations could not be detected for lack of enemy movement. Anchor communications / relays proved to be a must. Communications never failed with the newly inducted HX Sets.

The casualty evacuation of many men at the same time was a horrendous experience. Mountaineering techniques needed to be mastered. Constant counter insurgency commitment in past seven years had taken its toll on the Special Forces units. Expectations were very high – best or worst, if unaccomplished. Selfless commitment and sacrifice would be of a high order but the absence of the understanding of the Special Forces and special operations at various levels proved to be the bane of the problem.

These were operations conceived by people with dreams, men of ordinary fame but extraordinary courage and mental strength. These were people who would make a heap of all their earnings, pitch it with the flip of a coin and never breath a word about their loss. All these ordinary men worked beyond their capabilities and endurance, to satisfy a system which thought that a helicopter would be more expensive than all their lives. The partial success of this operation showed its great impact with enemy reinforcements coming by helicopters to stop the advance of Special Forces. Many succeeded only partially but even under enemy fire, enemy casualties in the far distance were being registered.

3

Op 'Black Tornado' - NSG Operations

In the middle of September 2008, the CIA station chief in the US Embassy in New Delhi called up the Joint Secretary (Liaison) in the external intelligence agency R&AW. The station chief passed on some intercepts that their Jalalabad station in Afghanistan had picked up. The intercepts indicated that the Lashkar-e-Toiba was preparing for a major terrorist attack in Mumbai against some key installations.

This was passed on to the liaison branch which immediately issued the alert to the Joint Intelligence Committee and the Multi Agency Centre of the Intelligence Bureau against the normal practice of processing raw intelligence before sending it out. Usually, any Pakistan related intelligence would go to the Pakistan Production desk of R&AW where the dots would be connected to present a comprehensive and actionable intelligence input. Instead, the input got lost in the maze of the intelligence bureaucracy.

On November 13, similar intelligence inputs began to surface with specifics like the Taj Hotel in Mumbai being described as a target for a possible terrorist attack. By November 19, another specific input about a "Lashkar-e-Toiba Ship" leaving the Karachi port in Pakistan was sent off to the various recipients including Naval Headquarters. But the War Room of the Navy's Operations' Directorate would get to look at the input only on November 22. By this time the "LeT Ship" had become an untraceable dot in the thousands of boats of various sizes that sailed the Arabian Sea waters for fishing.

Back in New Delhi, the National Security Guards (NSG) located at Manesar, a decrepit small industrial hub on the outskirts of the capital city was kept out of the loop. The Group Commander (Intelligence) those days was an officer on deputation from the Central Industrial Security Force (CISF) and had no training in gathering, let alone processing intelligence.

Raised in the aftermath of the assassination of Prime Minister Indira Gandhi, the NSG was supposed to be a Special Force organisation dedicated to carry out counter-hijack and hostage-rescue-intervention operations. The force would have two broad components. There would be two battalions, the 51 and 52 Special Action Groups (SAG) that would be exclusively staffed with volunteers from the Indian Army who would be supported by Special Ranger Group battalions which would draw its personnel from the central and state police volunteers. While the operations and training wings would be headed by Army officers, the overall force would come under a senior Indian Police Service (IPS) officer.

This led to several anomalies that were unique to India and would repeatedly prove detrimental to the NSG's operational capabilities. Most of the IPS officers being sent to command the NSG had never served with it at any point of their careers. Worse, many of them had very little experience in hostage rescue or counter-hijack operations, having spent their entire careers dealing with 'law and order' problems. On the night of November 26/27, when the call from the Cabinet Secretary came to the Director General of the NSG, the force was in a similar situation. The man heading the NSG was an IPS officer who had spent the better part of his life as an investigating officer in the Central Bureau of Investigations. He was known more for his association with the Bofors case and had no experience in any special operations or even basic hostage-rescue-intervention operations.

The Inspector General (Operations), a post reserved for a Major General from the Army was on leave and the Force Commander responsible for logistic support, also from IPS cadre, had little clue about what to do. The burden of mobilizing the NSG Task Force that would head to Mumbai fell on the Group Commander of the 51 SAG, Col Sunil Sheoran, the only Special Forces officer then on deputation from the Army in entire NSG. Manesar, nearly 35 km from the airport was a tactically inconvenient place to be. The distance would have to covered by road through excessive traffic and enroute one of the trucks broke down. This led to a major delay while the government was desperately searching for an aircraft that could take the NSG Task Force to Mumbai. After much searching they located a Russian IL-76 with the Aviation Research Centre, a wing of the R&AW that was available. Meanwhile, the Home Minister was merrily declaring on national television what number of NSG commandos was being dispatched to Mumbai and how many hours it would take for them to reach – all of which would have amused the terrorists in Mumbai as well their handlers in Pakistan no end.

By the time the Task Force arrived at the airport in New Delhi, the Group Commander was told that they would have to engage terrorists at one single installation. It was only when he was about to board the aircraft that he was told that he would have to simultaneously intervene at three places. With a Task Force with barely 200 commandos and support staff, this would prove to be a Herculean task.

Just a minutes before the Task Force took off, an Additional Secretary from R&AW was sent to accompany them but he did not discuss any inputs with the Group Commander. The team did not have anyone from the IB for any intelligence support even though intelligence agencies were plugged in and were monitoring the conversations between the terrorists and their handlers in Pakistan. None of these inputs were conveyed to the NSG Task Force when it landed in Mumbai at 0500 Hours.

Much of these systemic issues were germane to the way the NSG was forced to operate. They had faced similar issues when the NSG under the command of Col Jaideep Sengupta, another Special Forces officer and Sri Lanka veteran, had taken them to take on terrorists who had attacked the Akashardham Temple in Gujarat. With no intelligence support, Col Sengupta had to conduct a delicate hostage-rescue intervention operation over a two-day period.

The Task Force landed at the 0500 hours in Mumbai and received some cursory briefings from the Mumbai Police. The city police had lost some of its best senior officers in the initial hours of the assault and were still groping around in the dark for a semblance of command and control. The Task Force went to the State Government's Headquarters in

Mantralaya and met the Director General of Police who also had sketchy details to share with the Group Commander. By 0900 hours, the NSG was deployed at the three locations but were able to begin operations only at two of them.

It was then that they realised the mammoth task with little help from anyone else. The next Task Force would take off from Delhi and only arrive by later afternoon and time was running out. A small team of the Indian Navy's Special Forces, the MARCOS, had been hastily deployed at night to contain the situation and they engaged the terrorists at the Taj Heritage hotel for a few hours before the NSG arrived. The MARCOS had been exchanging fire from a static position and briefed the NSG after it was deployed. Major Sandeep Unnikrishnan would be part of the team that would be deployed at the Taj Hotel. With no building plans the small team had to hunt down the terrorists from the 1600 Rooms of the hotel that had two distinct wings. One wing had six storeys while the other one had 21 storeys. Intelligence indicated that there were four terrorists inside the Taj while there were two each at the Oberoi-Trident and Nariman House.

At 0920 hours, the NSG began operations and discovered that the terrorists had managed to evade the MARCOS using a stairway inside a kitchen that no one knew about. The commandos began a floor-by-floor intervention in a 60 hour operation that was now christened as *Operation Black Tornado*. With thick granite walls, rocket launcher fire was proving to be inadequate and there was a major paucity of snipers. With major limitations, the Group Commander fell back on innovation that would prove to be invaluable. Using a fire brigade's trolley he positioned his sniper who would start dominating the hotel from various angles as the trolley swivelled from one angle to the other. This gave the impression to the holed up terrorists that several snipers were dominating the situation and helped the commandos get a tactical advantage.

The commandos began a top-down approach so that they could clear out the rooms from a tactically-superior position. But in the absence of a master key each room was taking at least five minutes to clear out. The hotel staff had also suffered major casualties and very few people were available to aid the commandos. As operations continued, Major

Unnikrishnan's buddy commando suffered serious injuries when a terrorist fired at him. Major Unnikrishnan immediately evacuated his buddy and rushed up the stairs after the terrorist. This would prove to be fatal for him as the terrorist sent out a burst of automatic fire that would kill Major Unnikrishnan on the spot. He would be one of the many that night who would be posthumously awarded the *Ashok Chakra*, the country's highest peacetime gallantry award.

Saddled with a small task force, the NSG was struggling to localise the terrorists. Many of the trapped guests had locked themselves in their rooms complicating the rescue operations.

In the nearby Oberoi-Trident, the commandos had taken a similar top down approach to ensure a tactically superior position while they searched for the terrorists. The first fire fight would take place on the eighteenth floor in room number 1856 when a terrorist made a run for it towards the lifts. While a commando chased him the others broke into the room and lobbed a few grenades. The second terrorist was hiding in the bathroom and the commandos began to advance. Several grenades and bursts from their MP-5 sub-machine guns proved useless initially since the terrorist had taken cover inside the bathtub. Under withering fire the commandos managed to force their way into the bathroom and neutralise the terrorist. The other one had already been neutralised near the lift successfully.

At the Nariman House, reports had indicated the presence of an Israeli Jewish family that had been targeted by the terrorists as per their original plan. Much of the planning had received a lot of help from an American LeT terrorist, David Coleman Headley, who had visited Mumbai on six occasions prior to the attack to gather adequate intelligence while posing as a tourist. The Jewish house that ran a *Chabad House* from the premises was a key target for the LeT terrorists.

With such cheek-by-jowl constructions the NSG decided to rope in two Mi-17 helicopters from the Indian Air Force to rappel down to the terrace of Nariman House. The terrorists had blown up the staircase to prevent a counter attack and rappelling down to the terrace was the only option left to intervene. Without any armour plating the IAF pilots took off with the first batch of commandos to land them on the roof of Nariman

House. But the pilots and the NSG commandos had forgotten to exchange their radio frequencies. With no communication between the commandos in the air and their counterparts on the ground, the first helicopter hovered over the wrong roof as the first commando began to rappel down. Frantic hand signals from the commandos on the ground helped the pilots realise their mistake and they swerved off to make another sortie and head for the right roof.

This proved to be a major boon because the commandos surrounding Nariman House had seen a terrorist move towards the terrace when the first helicopter had hovered over the wrong roof. If the terrorist opened fire, it could have led to a major catastrophe for the commandos. The "mistake' proved to be a boon as the Mi-17 came back for the second sortie and the commandoes rappelled down to the roof. This operation would see the second major fatal casualty that the NSG would suffer. With the terrorists holed inside the room and the hostages killed nearly two days ago, one of the commandos, Havildar Gajendra Singh made the first attempt to enter the room. However, he would fall to a hail of bullets and succumb to his injuries. The NSG's room intervention expertise would be used to breach the wall using frame charges and the two terrorists would be neutralised by the NSG at Nariman House on the morning of November 29, nearly three days after the assault began.

In Taj, the last terrorist was still proving to be elusive when he was spotted by Colonel Sheoran. He would squeeze off a few shots that would take out the last surviving terrorist. The overall operation would be subjected to a barrage of criticism from the Special Forces community, many of whom had served in the NSG at some point. But the dilution of standards, the unavailability of quality Special Forces personnel and the unfamiliarity of the IG (Operations) with special operations had already taken its toll.

Many of the systemic issues that existed earlier and prevailed in 1999 continue to bog down the NSG even today. During the Air India's IC-814 hijack, the NSG that could have perhaps stopped the aircraft at Amritsar while refuelling was not moved. Again, the IG (Operations) was away and, the only psychologist posted was on leave and there was no dedicated aircraft available to move the Task Force. An ill-conceived plan to expand the NSG by opening new hubs across the country has led to a further dilution of standards. In fact, this has also led to shortages of basic weapons and ammunition that has had an adverse impact of training for specialised skills that are critical for the NSG to remain a Special Forces unit. But the absence of a centralised command and control for the Special Forces would prove to be a major stumbling block for the NSG as well.

4

Still Birth Special Forces Regiment and Special Forces Training School

In 1979, President Jimmy Carter's administration in the United States was presented with a major crisis. The Islamic revolution in Iran had stranded the employees in the American Embassy in Tehran and they were being held hostage. Negotiations had spun out of control and President Carter was desperate for a solution that could get the Embassy personnel back home safe.

After several rounds of deliberations President Carter agreed to an ambitious plan that had been put forward by the US Military. He was quickly briefed about a secretive military unit, known as the Delta Force, which was a part of the United States Army Special Forces and had been raised and trained for exactly the kind of a mission that had now come up.

The Delta Force was a popular name for Operation Detachment Delta, a new unit that was created after an American Special Forces officer returned after a stint with the British SAS. Major Charles Beckwith was so impressed at the vast range of skills and capabilities that the SAS had built up over the years, he felt that this was exactly the kind of unit that the American Army's Special Forces were looking for. Once the top military leaders were convinced about the need for such a unit, he started to raise a unit that would rival the SAS. He used the same structure, calling his teams "squadrons" just like the SAS described its company level sub units. Delta Force had not been deployed in any major operation and the planned mission in Iran would be its first.

Operation Eagle Claw proved to be a massive failure soon after it was launched. There were so many disparate elements and forces that no one really knew who was doing what. This led to a situation where the communication gear was not in sync and there were instances when the Special Forces men on the ground did not even share the same frequencies. With both sides "blind" to each other's capabilities, *Operation Eagle Claw* became a spectacular disaster as a helicopter and a C-130 fixed wing aircraft crashed in the Iranian deserts. The failure led to a lot of soul searching within the American military and the Joint Chiefs of Staff commissioned a Special Operations Review Group headed by Admiral James L Holloway with five other members to look into the causes of the failure. The Review Group produced a comprehensive report that would explain why the mission had failed and recommend major reorganisational changes.

But while the military was looking at fixing the problem, the United States Congress and the Senate was looking at a larger resolution. Two Senators would work closely with the military and other experts and push through the Goldwater-Nichols Reorganisation of the Pentagon Act that would mandate comprehensive military reforms. However, it was felt that the Goldwater-Nichols Act had failed to address the Special Operations issues comprehensively. So two other Senators, Sam Nunn and William Cohen, one Republican and the other a Democrat, got together and studied the problem. They introduced an amendment to the Goldwater-Nichols Act that was quickly approved by the United States Congress, as a consequence of which a separate Special Operations Command (SOCOM) at par with other existing Commands of the United States Military was established. The US Senators had recognised, at a political level, along with the senior military leadership that the Special Forces needed to be dealt with differently if politico-military operations were to succeed.

This needed a separate structure, command and control, doctrines, equipment, tasking, deployment capabilities and intelligence support. SOCOM would provide all this to the Special Forces and lead to a major synergy with the other Commands of the United States. What the two Senators and the Military leadership had done was a fact that was recognised by virtually every country in the western world. In the United Kingdom, the SAS had direct access to the Prime Minister and in recent years the Director SAS was reorganised and upgraded. A Major General was placed as the Director (Special Forces) instead of a Brigadier, and he was now in charge of all the Special Forces of all the three Services – Army, Navy and Air Force. In Israel, the secretive *Sayeret Matkal* reported directly to the Chief of Staff, and also ensured that it had direct access to the political leadership. In fact, Israel became the only nation in the world which had sent several Prime Ministers with a 'Special Forces' background to the *Knesset* (Israeli Parliament). Prime Minister Yitzhak Rabin had served in the precursor to the Israeli Special Forces known as the *Palmach*, followed by Prime Ministers Ehud Barak and Binyamin Netanyahu, who had all served and commanded the *Sayeret Matkal*. This displayed the kind of importance the Israeli political and military leaderships gave to the Special Forces and its potential to achieve strategic aims.

Back home in India, the Para Commando Battalions were making a mark in the annual formation level exercises. In one instance, a young Major (Prakash Katoch) displayed tremendous initiative in conducting a brilliant raid on an exercise Corps Headquarters in the wee hours of the morning, capturing the duty officer and whole set of maps and documents. He had his tall burly Second-in-Command (Captain I J S Sandhu) masquerade as a Divisional Commander and himself posing as his driver, drove into the Corps Headquarters with complete confidence, with flashing red beacons through the route meant for flag cars and accompanied by an officer from the exercise umpire organisation. In a subsequent formation level exercise, he, along with his assault team managed to infiltrate into an airbase maintaining complete surprise and conducted a highly successful raid. Operational experience was to come later in counter insurgency and counter terrorist operations in the northeast and later in J&K.

Unfortunately, in India, which needed its Special Forces more than ever, the Military leadership would continue to ignore the obvious and take several retrograde steps that rolled back the few progressive steps that had been taken to give the Special Forces their due. A decision was taken to raise a separate Special Forces Headquarters, Training Wing and a Special Aviation Unit. The original decision was to place all components at Banbasa to ensure a more centralised organisation. For operations in mountains, a team of one Special Forces unit and another team from a newly converted unit would be earmarked. For operations in plains and deserts, two teams of the newly converted unit and one team of a desert specialised Special Forces unit would be kept in readiness. For operations in the jungles and riverine terrain, two teams of the original desert specialist Special Forces battalion would be placed at the disposal of Army Headquarters.

It was also decided that manpower selection was to be from two to five years of service. A second tenure was permitted only for officers with at least nine to thirteen years of service. Most of these decisions were supported by Gen B C Joshi and his DGMO, Lieutenant General Vijay Oberoi. If the Special Forces community in India had a friend and a well-wisher akin to Lt Gen Nanvatty, it was Lt Gen Oberoi. Commissioned into the Maratha Regiment, Lt Gen Oberoi had overcome a debilitating disability when he had to lose a limb after being grievously injured in operations. However, he managed to recover and rapidly rose through the military hierarchy to retire as the Vice Chief of Army Staff. Lt Gen Oberoi was one of the few officers who recognised the need for a Special Forces as had been recommended by the Special Forces study group in 1988. He also had the distinction of being the first and only Colonel of the Special Forces Regiment before this fledgling Regiment was merged back into the Parachute Regiment.

When the need to raise a fourth Special Forces battalion rose, the Parachute Regiment, which had been the traditional source for Special Forces battalions, declined because now there was a separate Special Forces Regiment. Lt Gen Oberoi, as the Colonel of the Maratha Light Infantry Regiment spoke to the various Commanding Officers and offered one of the most distinguished battalions of the Maratha Light Infantry Regiment for conversion into Special Forces role.

While the new unit began its conversion in Nahan, a small hill station in the state of Himachal Pradesh, a new Special Forces Headquarters was being raised. The existing three Special Forces Battalions and the fourth unit undergoing conversion had already been pulled out of the Parachute Regiment under Gen B C Joshi's orders and set up in a Special Forces Regiment for the first time.

On a summer day, while Colonel Jude L Cruz, one of the Commanding Officers of a Special Forces Battalion was visiting Army Headquarters in the historical South Block, he was hastily called in to attend a special ceremony. It was ironical that one of the most significant moments in the history of the Indian Army would be allowed to pass of with such a routine manner.

The ceremony was of the Banner of the newly raised Special Forces Regiment to be presented to General Shankar Roychowdhury, who had taken over as the Chief of Army Staff when General B C Joshi suddenly succumbed to a massive heart attack while still in service. The presentation of the banner was to be made by Lt Gen A Sandhu, the then Director General of Military Operations. The DGMO had been nominated as the ex-officio Colonel of The Special Forces Regiment to ensure that it received patronage at the highest levels of the military. Similarly the Chief of Army Staff would become the Honorary Colonel of the Special Forces Regiment. So, on a summer day in 1996, Colonel J L Cruz accompanied Brigadier Keshav Padha, the newly appointed Commander of Headquarters Special Forces and Colonel Anil Verma, Director of the Special Forces Cell in the Military Operations Directorate (both veteran Special Forces officers) to present the Banner to the Chief.

Unfortunately the appointment of Honorary Colonel of the Special Forces Regiment was to be held by the Chief for a few months only as the Special Forces Regiment was later scrapped under extreme pressure by the Parachute Regiment veterans. Brig (later Major General) Padha fought a lone battle to ward off the attacks and the DGMO had fully supported him but the Chief could not withstand the pressure and gave in to the demands of the Parachute Regiment veterans. Later on, while explaining this retrograde decision he would call Brig Keshav Padha to his office and mention that the "*Gaon Boodas* (Village elders) from your Regiment had virtually blown up the roof top of my office till such time I agreed to scrap this new Special Forces Regiment." Frankly, Brig Padha was frustrated at this retrograde action on the Chief's part but could do little to take on this powerful antiquated clique of tradition-steeped officers and veterans.

In a single stroke, Gen Shankar Roychowdhury as the Army Chief set back the Army's Special Forces by several decades. All the work and the sacrifices that had been accomplished by his predecessors, the Brig N Bahri Committee and the Special Forces themselves was allowed to be finished off in a single moment of weakness. The political leadership was blissfully unaware of the Army Chief taking such a major decision of disbanding a Regiment and merging the Special Forces Battalions with those in the Parachute Regiment. The fact that Special Forces and the airborne forces are as different as the Infantry is from the Armoured Corps or the service corps was ignored and misplaced tradition chosen over pragmatism and professionalism. Worse, a move that would have met India's strategic needs was sacrificed at the altar of the egos of a few disgruntled army veterans.

Around the time that the Special Forces Regiment was being created, Army Headquarters had also begun preparations on another major initiative for the Special Forces. Raised in 1993 as the Special Training Wing (SFTW), this organisation was established to train the Army' Parachute (Special Force) personnel. It functioned in tandem with Headquarters Special Forces including for conducting probation training for personnel posted to PARA (Special Force) till HQ Special Forces was disbanded. Subsequently, it also catered to extraneous requirements of various headquarters and formations from time to time ranging from training of Ghatak (Commando) platoons of infantry battalions to even amazingly running security capsules for personnel below officer rank proceeding on pension. In 1998, the role of the SFTW was reviewed and the organisation was proposed to be reduced by some 19 all ranks. Prior to Operation Parakaram, the establishment was training approximately 400 personnel on yearly average through various courses. This included Aqua Adventure Capsules on behalf of the Army Adventure Wing.

In 2002, it was found that the SFTW was not able to meet the advanced specialist training requirements of the PARA (Special Force) personnel in full. An individual was then going to the SFTW for training on various courses after he had been absorbed in the PARA (SF) battalion. At SFTW he received advanced training on his primary trade (demolitions, communications, specialist weapons, medical or navigation), basic exposure to which he had already received during probation in his unit. Then there was also refresher training through courses at SFTW of special skills like under water diving, combat free-fall, missiles etc, having acquired the skill initially at other schools of instruction.

The then role of SFTW did not cater to all revised tasks of PARA (Special Forces) like reconnaissance, surveillance and target designation. No Language training was being conducted at SFTW as there were no language instructors, and this was left to the very limited capacity of the Army Education School & Centre rather than organising it at the School of Foreign Languages. Though training of Indian Air Force (IAF) personnel was part of SFTW's role, IAF then had no Special Forces. An examination indicated that the SFTW needed to run 10 types of courses with varying periodicity in a year, not counting the Aqua Adventure Capsules run under aegis Army Adventure Wing. Considering the average yearly intake of then five PARA (Special Forces) battalions, SFTW needed to train a total of approximately 1200 personnel from the five PARA (Special Forces) battalions including refresher training in order to maintain these units at peak efficiency. It was therefore decided to cancel the reduction of manpower as per the 1998 review, increase the establishment by 27, change its name from SFTW to Special Forces Training School (SFTS) and accord it the status of a Class 'A' training establishment in order to post best quality instructors and attenuated facilities for imparting better training.

In its present temporary location, also organising joint training with foreign Special Forces, the SFTS has only basic training facilities and even a firing range only up to 600 metres. Existing infrastructure for trainees does not cater for the advanced Special Forces training to be imparted to the required number of personnel. More significantly, the number of PARA (Special Forces) Battalions in the Army has increased from five in 2002 to now eight and two more are to be added. Move of SFTS to its permanent location at Bilaspur in Chhattisgarh was approved long since but is languishing for want of acquisition of land by the State Government. An early shift of SFTS is urgently required as creation of required training facilities like firing ranges (long and short ranges, indoor shooting ranges, room intervention range, combat shooting range, sniper range, jungle lane shooting range), underwater diving tank, underwater demolition tanks, training model rooms, rock climbing walls / areas, obstacle courses, wind tunnel , hangars for ground and air maneuver training, helipads, drop zones, laboratories, gymnasium and work stations, swimming pool, unarmed combat area and the like, plus the administrative set up required all have considerable gestation period in establishing and becoming fully functional.

Aside from the advanced training already being conducted in basic traits of Special Forces personnel as mentioned earlier and Combat Leaders Course, training should also be conducted in operational techniques and skills like intelligence collection, interrogation, tracking, sniping, disguise, training with dogs akin to the USSF etc. Refresher training in combat free fall also needs to be initiated after move to the new location. In its new location, the SFTS needs to become a 'centre of excellence' for the Special Forces of the India Military.

Considering the current proposal to integrate the Military Special Forces under a Joint Command, it would be prudent to convert the SFTS into a Tri-Service Special Forces training establishment (once it moves to its permanent location in Chhattisgarh) with the existing Naval Special Warfare Tactical Training Centre as an adjunct. This would be a vital step towards integration of the Military Special Forces. The command and control of the SFTS has been fluctuating between Directorate General of Military Operations, Directorate General of Infantry, Headquarters Western Command and Headquarters Army Training Command. In keeping with future requirements, it would be prudent to convert the SFTS to a Tri-Service establishment (as mentioned earlier), rename it as MSFTS (Military Special Forces Training School) and place it, along with its adjunct Naval Special Warfare Tactical Training Centre directly under the proposed Tri-Service Special Forces Command / Commando Command.

5

Special Forces and Airborne Forces

"I find the vision blurring in certain quarters on the issue of Parachute and Parachute (Special Forces) units. I am very clear that a Parachute Battalion is simply an infantry battalion in airborne role and has nothing in common with a Special Forces Battalion. Also, the Special Forces are not a game of numbers and I for one am against their expansion of any sort. Our Special Forces in their present state are comparable to the Rangers of USA. We must consolidate and modernise our existing Special Forces resources. As regards the Parachute Brigade, I view them as a rapid reaction force to be used within and outside the country".

- Lt Gen RK Nanavatty, Army Commander, Northern Command (2002).

Special Forces, as their appellation indicates, are 'special' and that by itself should drive home the point that they are of a strategic nature primarily tasked for employment beyond national borders – a logical term of reference that is usually lost in the fog of wooly and hasty planning, mostly due to ignorance and inability to grasp the strategic environment, its setting and compulsions under which such forces are employed. There are many definitions for Special Forces, one of which says, "Special Forces are a term used to describe relatively small military units raised and trained for reconnaissance, unconventional warfare, and special operations. These exclusive units rely on stealth, speed, self reliance and close teamwork, and highly specialised equipment. Traditionally, the mission of the Special Forces are in five areas; counter-terrorism, unconventional warfare, facilitating the internal defense of foreign countries, special reconnaissance and direct action against specific targets.

Who does not want to belong to or feign he is from Special Forces? Two decades back you could spot sloppy security guards outside shops in Connaught Place, New Delhi sporting 'Commando' or 'Special Forces' on their shoulders with maroon berets worn at rakish angels. It is so very convenient where little action is taken against anyone wearing military insignia. The Central Armed Police Forces (CAPF) and Police in any case wear the same badges of rank, something you do not see in other countries that help them seek parity of pay and perks with the Military. Today an officer in the Sashastra Seema Bal (SSB) becomes a Deputy Inspector General (DIG) with seven years service, enjoying the pay of a Brigadier / equivalent rank officer of the military. Wikipedia showcases some 50 odd Special Forces of India – ranging from Army Special Forces, Marine Commandos (MARCOS) of Navy, Garud Commando Force of Air Force, Special Action Groups (SAGs) of National Security Guard (NSG) to Special Protection Group, Special Frontier Force, Cobra, Railway Protection Force Commandos, Quick Reaction Team of Indian Railway, Special Task Forces in each State, Special Operations Groups of J&K and Rajasthan, Mumbai's Anti Terrorist Squad and Force One, Grey Hound Commandos and Octopus of Andhra, Straco and Combat Force of West Bengal, Jaguar Force, Chhattisgarh Commando Battalion (CCB), State Security Guards, Anti-Guerilla Force and what have you. Little wonder then that our media is a confused – calling everyone 'elite' and 'Special Forces'. Ironically, there is difficulty, even in some cross sections within the military in differentiating between 'Special Forces' and 'Airborne Forces', which is both advertent and inadvertent.

Ambiguity of the term 'Special Forces' within the Army comes up periodically, facilitating categorisation of non-Special Forces into the Special Forces category. Such pseudo specialisation harms both ways – sidetracking implementation of the Concept of Special Forces in the true sense and distracting non-Special Forces units from their primary task. We are the only Army in the world where Parachute (Special Forces), shortened version being PARA (SF) units and regular Parachute, shortened version being PARA units are clubbed into the same Regiment; The Parachute Regiment. This aberration happened despite the fact that the first two Special Forces units 9 and 10 PARA (SF) did not draw manpower exclusively from The Parachute Regiment. Grouping the PARA (SF) and PARA units in the same Regiment resulted in the latter continuously seeking parity for obtaining the Special Forces allowance and their insignia. The Army realised the folly in early 1990s and took steps to rectify this anomaly. A separate Special Forces Regiment was created in 1994 clubbing the then three Special Forces units – 1, 9 and 10 PARA (SF). Major General (later Lieutenant General) Vijay Oberoi, then Director General of Military Operations was appointed Colonel of The Special Forces Regiment. The three units were renamed 1 SF, 9 SF and 10 SF. A decision was taken to raise Headquarters Special Forces in order to have a central agency oversee strategic tasking, operational employment, intelligence inputs, capacity building, manning, equipping, training and consolidation of Special Forces. In addition, the appointment of Deputy Director General Military Operations (Special Forces) was sanctioned in Army Headquarters as DDGMO (SF). Unfortunately, the then Chief of Army Staff, General BC Joshi died in harness and within a short span of the raising of The Special Forces Regiment and General Oberoi was posted out of Army Headquarters. This gave an opportunity to a host of retired and senior level PARA officers to collectively tackle the new Army Chief to reverse the decision on the plea that The Parachute Regiment had been 'broken'. No views of the then existing Special Forces officers were sought on the issue, who all wanted the Special Forces Regiment to continue. This apart, the following steps were taken to prevent re-creation of a Special Forces Regiment in future: raising and location of Headquarters Special Forces was done at a remote location like Nahan instead of Delhi (as earlier intended) so that it remained ineffective. This headquarters was later merged with Infantry Directorate and finally disbanded with its appointments merged with Military Operations, Infantry and Weapons & Equipment Directorates; and the appointment of DDGMO (SF) in Military Operations Directorate was made tenable by both PARA (SF) and PARA officers.

The Army now had a situation in that PARA officers without having any Special Forces experience, rendered advice on Special Forces issues from a chair of authority; as is the case even today. This situation had the following fallouts: 9 and 10 PARA (SF) who specialised in mountains and deserts respectively were moved out to Jodhpur and Agra respectively despite protests by both Commanding Officers. The anomaly was finally corrected in 2001 by reverting them to parent locations; special equipment imported for the then three Special Forces units (1, 9 & 10 PARA (SF)) during 1984-85 was never introduced into Service since concerned cells in Infantry Directorate and Weapons & Equipment Directorate were not manned by PARA (SF) officers. This resulted in inability of the Master General of Ordnance Branch to provide any replacements in future; the appointment of Commandant & Chief Instructor, Special Forces Training Wing (SFTW), later renamed Special Forces Training School (SFTS) earlier always held by a PARA (SF) officer, was made tenable by a PARA / PARA (SF) officer while reviewing the WE of SFTW; probation for both PARA and PARA (SF) units was brought at par to 90 days and split between PARA Centre and concerned unit for 45 days each, overlooking protests by Commanding Officers of Special Forces Battalions. Concurrently, the Parachute Regiment (ordinary paratroopers) started pushing for converting the entire Parachute Regiment to Special Forces spearheaded by non-Special Forces Colonels of The Parachute Regiment–which continues todate.

The neglect of the Special Forces led to such extent that a Special Forces Brigadier (Prakash Katoch) while attending the National Defence College Course in year 2000 wrote a demi-official letter to the Chief (Designate), Lt Gen S Padmanbham, recommending that when he assumed the reins of the Indian Army, he should consider 'disbanding' the Special Forces because of the continued pathetic state of their neglect. As a result, in the year 2001, Gen Padmanabhan as the Army Chief ordered a major study for Modernisation of Special Forces that was to be headed by Brig Prakash Katoch who by then had been posted to the Military Operations Directorate. As a result of this study, major restructuring of PARA (SF) Battalions was done on tailor-made theatre specific units and a fourth assault team was added to each unit in addition to equipping these units with state-of-the-art weapons and equipment required for enhancing capabilities for various Special Forces missions. Probation period in PARA (SF) units was enhanced from three months to six months in keeping with operational requirements. Modernisation and equipping of the then five PARA (SF) Battalions was approved at a cost of approximately Rupees 400 crores and raising of an Army Aviation Special Operations Squadron was also approved with additional dedicated budget of Rupees 680 crores.

Most significantly, a conscious decision was taken for no further expansion of Special Forces till end of the 10th Army Plan, after which a review was to be undertaken. However, this decision was reversed within six months. The Colonel of the Parachute Regiment, then heading Perspective Planning Directorate at Army Headquarters used the US invasion in Iraq as an excuse in propagating that US had deployed some 20,000 Special Forces in that country. The aim was to convert more PARA units to PARA (SF). This was despite the fact that intelligence reports spoke of only "small detachments" of Special Forces having been employed by the US, bulk of which were inducted into Northern Iraq a year and half before the invasion. Propagation of the figure 20,000 included 82 and 101 Airborne Divisions of the US, which are not Special Forces and volunteers from these formations for US Special Forces have to undergo a full-fledged probation before they can join Special Forces. Incidentally, of the present 60,000 strong US Special Forces Command (SOCOM) with a [1] 2013 budget request of $10.409 billion, the real Special Forces element is only 15,000 strong of which Psychological Operations Teams and Civil Affairs Teams are not uniformed fighting men. Even during peak period of Special Forces deployment in Iraq, only 90 x Operation Detachments Alphas (ODAs) were actually used (each ODA is 10-12 men strong). Once the decision of no expansion was reversed, orders were issued to convert another two PARA battalions to Special Forces. With conversion of these two PARA battalions to Special Forces, plus six assault teams added as a result of the Study on Modernisation of Special Forces, the Indian Army went in for addition of four and a half battalions worth of Special Forces in a span of just three and a half years.

This was in complete disregard to the four universally acknowledged[2] Special Forces Truths: humans are more important than hardware; quality

───────────────────────────────

[1] http://www.fas.org/sgp/crs/natsec/RS21408.pdf

[2] http://www.soc.mil/USASOC%20Headquarters/SOF%20Truths.html

is better than quantity; Special Forces cannot be mass produced; and competent Special Forces cannot be created after emergencies arise.

Expansion of Special Forces in foreign armies is very deliberate. Authorised expansion rate of US SOCOM is 1.8 percent annually. However, Admiral Olsen (erstwhile Commander US SOCOM bid for 2.5% expansion in 2011 and Admiral Mc Raven, present Commander US SOCOM bid for addition of 3000 which includes 'support elements' due to increased responsibilities. Pakistan added the fourth SSG battalion only in recent years. In case of our Army Special Forces, we went in for a 120% increase in period 2001-2004 alone including converting three PARA battalions to Special Forces (a third one was converted in addition to the two mentioned above) and adding the fourth assault team in all Special Forces units. Ironically, validation of newly converted units was left to tenure in counter insurgency environment. Concurrently, little effort was made to enlarge the advanced Special Forces skills training capacity of the Special Forces Training School (SFTS). Though a review was undertaken in 2002, the decisions for expanding the capacity, using it exclusively for advanced training of Special Forces, appointment of Commandant & Chief Instructor to be exclusive to Special Forces officers remained in limbo. Rapid expansion adversely affected the manning, equipping and training of these units, more importantly the combat potential.

Manning becomes problematic when there are restrictions on the Commanding Officer of Special Forces units in terms of annual percentages in respect of wastages. Many a time, the personnel who report for probation are sub-standard and units have to work overtime to get them in shape or shunt them out only when individual cases are approved by the Infantry Directorate at Army Headquarters. This implies delay in months with ample chance of non-acceptance on plea of authorised annual percentage. Significantly, in the US Special Forces, the rejection rate of volunteers from even US Airborne Forces is 70 – 75%. In Russia's Spetsnaz it is 80%. Our own MARCOS has a rejection rate of 60 %.

Perhaps we could learn something from the selection process of foreign Special Forces. In the Special Services Group (SSG) of Pakistan, an individual first undergoes a 24 week initial testing and training process followed by a nine month training-cum-selection process in all types of terrain, including insertion and extraction. It is only at the end of this 24 weeks plus nine months that the individual is accepted into the SSG. It may be recalled that the Brig. N Bahri committee had also recommended a probation cum continuation training of one year duration. In the US Special Forces, the Special Forces Qualification Course (SFQC) or informally, the Q Course is the initial formal training program for entry into the Army Special Forces. Phase I of the Q Course is known as Special Forces Assessment and Selection (SFAS). Getting "Selected" at SFAS will enable a candidate to continue on to the next of the four phases - which if a candidate successfully completes he will graduate as a Special Forces soldier and will generally be assigned to a 12-man Operational Detachment "A" (ODA), commonly known as an "A team." The length of the Q Course changes depending on the applicant's primary job field within Special Forces and their assigned foreign language capability but will usually last between 56 to 95 weeks.

In 2003, another study on Special Forces was ordered but it was shelved when General Nanavatty (erstwhile Army Commander, Northern Command) pointed out that that there was no requirement of another study when all the issues related to Special Forces had already been addressed by studies conducted earlier including a major one done in 1980s, latter having led to the formation of a Special Forces Regiment. The proposal was therefore dropped.

But post retirement of Gen Nanavatty, yet another study on Special Forces was ordered in 2004 headed by a regular Paratrooper and not a Special Forces officer. This study made some hilarious recommendations like renaming of the PARA Battalions to Parachute Special Forces (Airborne), authorising them the Special Forces Insignia and Special Forces Allowance. The logic given was that PARA and PARA (SF) are the same force since both operate behind enemy lines. The study recommended that the Special Forces be expanded to 13 battalions by year 2010, each PARA unit to have minimum of 30 personnel qualified in Combat Free Fall, and the Special Forces Training School be merged and put under command The Parachute Regiment Training Centre, besides other recommendations. Suffice to mention, the Special Forces officer members refused to sign the study. There was a separate proposal that 30 Pathfinders in each PARA Battalion be based on Para Motors and this mode of transportation be optimally utilized in counter insurgency areas like in Jammu and Kashmir, till someone remarked in a discussion on the issue this would also require authorisation of bullet proof bum pads least we accept gunshot wounds on the backsides.

The CFF capability, as mentioned above, was authorised to the PARA battalions eventually but talking of CFF capability, the US maintains a CFF team of 135 personnel, who undertake 10 operational jumps by night every month with full combat loads including weapons and two rucksacks. If we are looking at creating 30 CFF qualified personnel in each PARA battalion who will probably be able to do one refresher course of hopefully 10 jumps in a period of one or two years, then we will end up with a lot of jumping jacks wearing the CFF Basic CFF Badge but sans required operational capability. Even the Para Motor capability introduced in PARA Battalions over the past eight years has resulted in a small number of personnel taking part in demonstrations. Using pathfinders en masse using Para Motors is an absurd concept to say the least, their only use being stealthy insertion in ones or twos provided the terrain and winds are favourable.

In recent years, the appointment of DDGMO (Special Forces) in Military Operations was elevated to Major General rank, as ADGMO (Special Forces) but the incumbent posted to date is still a non-Special Forces officer. The elevation occurred as a result of a detailed proposal sent by Lt Gen HS Lidder, then heading Headquarters Integrated Defence Staff (IDS) to the then Army Chief in 2008 to create a two star rank post directly under the Vice Chief of Army Staff with appropriate staff to oversee all matters related to Special Forces. Lt Gen Lidder, a highly decorated Special Forces officer had commanded a Special Forces Battalion very successfully in Sri Lanka under the Indian Peace Keeping Force. At the time of sending the proposal he was also the Colonel of The Parachute Regiment. However, this was seen as loss of turf by the Military Operations Directorate, and hence the appointment of DDGMO (Special Forces) was elevated in situ to ADGMO (Special Forces) in Military Operations Directorate with everything else status quo and continues as such. Ironically, before posting of the first AGMO (Special Forces), the appointment of DDGMO (Special Forces) was being held by a normal infantry officer who by his own admission had no idea of Special Forces or Special Operations. His only qualification was that he belonged to the same Regiment as the then Chief who wanted to give him the stamp of having served in Military Operations Directorate. That being the consideration of the Army to Special Forces no surprises to the ambiguity and understanding of the issue.

A peculiar situation continues in our Army, wherein, the PARA (Special Forces) and regular PARA units are clubbed into the same Regiment. Senior Paratrooper officers, who have never served in Special Forces, holding appointments dealing with Special Forces issues endeavor to establish parity between the tasking of Special Forces and normal Parachute units. Reorganisation of the PARA units on the same lines has been attempted in the past but has been shot down on valid grounds that the operational requirement of a Parachute Brigade and Parachute Battalions is undeniable. Similarly, The Parachute Regiment has been asking for the Special Forces Allowance also to be admissible to PARA units including during projections for the last Pay Commission but this was rightly not accepted by the Chief of the Army Staff. During the 2009 Reunion of The Parachute Regiment held at Agra in 2009, serving and retired PARA officers got after the COAS, Gen Deepak Kapoor that more PARA battalions be converted to Special Forces role so that they can get the Special Forces Allowance and the Balidan Badge (Special Forces insignia). The COAS responded that this will imply raising more PARA battalions who will start clamoring for conversion to Special Forces moment their raising is completed.

More recently, a proposal was again thrown up in continuation of earlier efforts to somehow gain parity between the PARA and the PARA (Special Forces). The only difference this time was that there is no mention of the Special Forces Allowance and the Balidan (Special Forces Badge) to PARA units since it was perceived these would automatically come in if the proposal was accepted. The proposal recommended that the PARA units be assigned the roles of 'Independent Small Team Actions', 'Guerilla Warfare', 'Sub-Conventional Operations in Unconventional Scenario' and 'Hostage Rescue', significantly, discarding their primary role of holding ground. With such extraneous issues coming up periodically, the Special Forces Concept gets side tracked. Who one may ask will then do the role of the Parachute Brigade and PARA Battalions? The fact that Parachute units, who are Infantry units with airborne capability, must continue to be mandated with tasks that is in support of a formation in their ground holding role. A PARA battalion in no way can be equated to a Special Forces unit. Foreign armies are very clear on the issue. The US has separate manning and training policies for Special Forces and Airborne Forces as well as separate training institutions for each of these forces.

Conversion of the Parachute Regiment's Parachute Battalions into Special Forces unbalanced the only strategic formation of the Army – the Parachute Brigade (now being addressed by raising more PARA battalions). De facto, it meant converting the Army's lone strategic reach capable and rapid reaction capable formation - the Independent Parachute Brigade. This approach appears taken on perception that there is no longer a threat that calls for a strategic intervention or rapid reaction capability by AB (airborne) troops as executed by the Parachute Brigade in Operation Vijay, Goa in 1961, Sikkim in 1965, Dhaka in 1971, Operation Cactus, Maldives in 1988 - a role that Special Force units are neither structured, trained, equipped or armed to perform, particularly if the intervention involves conventional attack, capture and holding ground for a length of time.

Therefore, conversion should have been thought of in light of past, present and future events since Gulf War II and India's strategic vision in the Indian Ocean Region from Central Asia to Indonesia and from the Horn of Africa to SE Asia and China. It is not the odd whimsy that armies of USA, UK, France, China, Israel and Russia continue to retain conventional airborne formations as distinct entities in addition to their Special Forces. Rationale justifying utility of airborne forces is available from battle reports of Iraq during Gulf War I and Gulf War II where low and high intensity warfare was embedded with terrorism. Airborne troops of Coalition Forces played a vital role in these operations in conjunction with USSF and SAS. Russian and British airborne forces have operated in Chechnya and Kosovo jointly with Spetsnaz and SAS respectively. Airborne forces of UK, France and US have performed classical joint airborne-Special Forces tasks (strategic and tactical) in Afghanistan and in Gulf War I and II as follows: force projection (airborne action) over land and sea into Afghanistan and Iraq – Kurdistan; "threat in being" to tie down / destroy Taliban elements at tactical / theatre level; capture and hold strategic ground in Kuwait and behind Iraqi lines through airborne assault; overfly anthrax contaminated Kurdish areas of Iraq to establish firm base by airborne assault / ground infiltration; established beachheads in conjunction US Marines in Persian Gulf; capture / recapture islands in Persian Gulf and Gulf of Hormuz by joint airborne / sea borne assault; operations in conjunction SAS / Delta Force to track and destroy mobile SCUD launchers; assisted UN peacekeeping operations in Somalia, Serbia, Bosnia and Kosovo; clandestine surveillance of WMD sites and manufacturing facilities. These tasks could not have been done by SAS or Delta Force on their own.

It is not that other armies have not faced friction between the Special Forces and those that are not but gradually they have overcome the same and established clear dividing lines.[3] Col Aaron Bank in his book "From OSS to Green Berets writes, "It was during the preparatory period that Rangers and Airborne members of the training staff working under Special Forces veterans gradually discovered how much more complex and complicated Special Forces training was compared to what they were used to. It was also apparent that Special Forces could conduct any type of behind-the-lines operations...... Special Forces are an outfit that conducts all aspects of unconventional warfare with emphasis on developing large indigenous guerilla forces".

Two decades back 82nd and 101 Airborne did their best not to let USSF like Delta Force come up. Indian Army is undergoing the same – even in greater intensity as Special Forces and airborne forces share the same Regiment. This has resulted in Parachute Battalions converting to Special Forces and raising of more Para Battalions who will continue to clamour for conversion to Special Forces. The Army needs to take corrective action. The requirement of the Parachute Brigade and PARA Battalions is loud and clear while we prepare for a combined China-Pakistan threat. You can hardly go any distance in mountains unless you establish airheads and go for rapid buildup in enemy rear including follow up air transported operations. The PARA Battalions must continue to concentrate on their primary role of ground holding. The mad race of PARA battalions for finding parity with PARA (SF) must be put at rest even if it implies de-linking them from PARA (SF). Additionally, the Army should ensure that appointments dealing with Special Forces issues must be held only by Special Forces officers.

---

[3] Bank, Aaron, From OSS to Green Beret – The Birth of Special Forces, Presidio Publishers, USA, 1986.

Lower rank officers officiating in say appointment of ADGMO (Special Forces) should be acceptable in case a two star Special Forces officer is not available.

Today's asymmetric wars are laced with unprecedented treachery, deceit and denial. On the question of pro-active employment of Special Forces, the fear of being labeled aggressor is fallacious since coping with non-traditional challenges does not equate automatically to physical attack. Special Forces provide us the tools to address non-traditional challenges to our security by providing a silent but effective medium. We need to develop the necessary will to contend with emerging strategic challenges. Their tasking should include asymmetric warfare, unconventional / fourth generation warfare, special operations, strategic reconnaissance, psychological operations and the like. We need to get a handle on the fault lines of our adversaries in order to achieve requisite deterrence. There is a need to go pro-active on the issue least we permit our economy and security to be weakened. We should not let this potent capacity be sidetracked by creating more and more pseudo Special Forces on advice of pseudo specialists.

6

Existing Status - Indian Special Forces

"India's Special Forces are ill equipped and underutilized."

- Gen S. Padmanabhan, COAS (2001).

The major part of the Special Forces comes from the Army; the Parachute (Special Forces) battalions, eight in number at present. They have been effectively employed in counter insurgency and counter terrorist operations, have considerable battle experience and are doing periodic joint training with foreign Special Forces of many countries. Next come the MARCOS numbering less than 1000 at present but is expected to double the number at a future date. Ironically, a case taken up by the Navy to raise a Marine Brigade has been languishing with the government for more than a decade. MARCOS too have had battle experience including under the Indian Peace Keeping Force in Sri Lanka and in counter insurgency and counter terrorism in India. The latest to join the Military Special Forces are the Garuds of Air Force. Established in 2004 and numbering between 1500 and 2000, they are yet to be blooded in operations though some are deployed in Congo under the UN flag. They have been undertaking protection of critical Air Force bases and installations, search and rescue during peace and hostilities and disaster relief during calamities. During hostilities, some of their tasks overlap those of their army counterparts.

The National Security Guard [1](NSG) was created in 1986 in the

---

[1] National Security Guards, http://www.bharat-rakshak.com/LAND-FORCES/Special-Forces/NSG.html

aftermath of 'Op Blue Star'. It operates under the Ministry of Home Affairs (MHA) and is primarily tasked for counter terrorism and anti-hijack, other tasks being bomb disposal (search, detection and neutralising bombs / IEDs), PBI (Post Blast Investigation) and hostage rescue. They have played a pivotal role in internal security and have undertaken a number of operations with numerous successes to their credit despite situations like Black thunder II where 1500 NSG personnel were pitted against 50 Sikh terrorists. NSG was the main Special Force employed during the 2008 Mumbai terrorist attack. [2]At the time of raising and for a few years thereafter, procurement of weapons, equipment and technology for the NSG had top priority and they could pick these up off the shelf globally. The first IG (Operations), Maj Gen Naresh Kumar could take an aircraft and pick up any weapon or equipment off the shelf in any part of the world. He had honed the NSG to optimum level of combat efficiency. The first ever live anti-hijack exercise conducted by him was so well executed that Prime Minister Rajiv Gandhi who was witnessing the exercise, made sure he was conferred with a second Param Vashisht Seva Medal. Maj Gen Naresh Kumar continues to date to be the only recipient of Bar to PVSM. In later years, procurement process of the NSG became as bureaucratic as the Military. This was noticeable during the 2008 Mumbai terrorist attack with total absence of 'corner shots' and deficient night vision equipment. The urgency has picked up post this operation but the rapid increase in manpower and attendant equipping needs to be coped with now. Additional NSG hubs have come up at Mumbai and Kolkata already and similar hubs are being set up at Chennai and Hyderabad. The cutting edge is the SAGs (Special Action Groups), the SRGs (Special Ranger Groups) having the task of providing the cordon. All the personnel in the SAGs and some support units, training and headquarters are on deputation from Indian Armed Forces, the rest being drawn from the central police organisations. The SAG is the offensive wing drawn from units of the Indian Army. The SRG consists of members from Central Armed Police Forces (CAPF) and Central Police Organisations (CPOs). Personnel of Special action Groups have been training with foreign counterparts in numerous countries including Israel, France and Germany.

---

[2] Kasturi, Bhashyam, National Security Guards: Organisation, Operations and Future Orientations, *Indian Defence Review*, Vol. 8 (3), October 1993, pp. 59-63.

There have been several occasions where the lack of proper transportation has hampered the response time of the unit. This was particularly evident during the 1999 hijacking of Indian Airlines Flight 814. Even during the 2008 Mumbai terrorist attack, the unit was delayed due to lack of aircraft in Delhi and then lack of ground transportation in Mumbai. The NSG [3] is ladled with numerous problem areas; the core of the problem is the attitude of the national security establishment to the employment of Special Forces. The NSG operates under the Ministry of Home Affairs (MHA) and the Director General is a Police officer who is just picked up from any State from Central Police cadre. In so doing, the very nature, orientation and character of the force has changed. Enroute to counter the 2008 Mumbai terrorist attack, the then Director General told the Task Force that he wanted the terrorists 'alive'; imposing undue caution and which perhaps was one of the reason for the time taken. The other problem is the lack of integral assets for a force of the NSG type, which is meant to possess rapid mobility, firepower and technology. The force does not have the wherewithal to generate its own intelligence too. Then is the continuous problem of a large number of personnel of the NSG tasked for protection of the country's VIPs (some two scores in number) despite a separate Special Protection Group (SPG) authorised for VIP protection duties and number of NSG personnel serving on deputation with the SPG.

The Special Frontier Force[4] (SFF) is a paramilitary force of India. It was conceived in the post 1962 Sino-Indian war as a guerrilla force composed mainly of Tibetan refugees. Based in Uttrakhand and also known as establishment 22, this force operates directly under the Cabinet Secretariat. SFF is headed by the Inspector General (IG) who works under the supervision of Director General (DG) Security, Research & Analysis Wing (R&AW) who reports directly to the Director of R&AW. The current SFF force levels are around 10,000 men. Individual battalions have a strength of around 900, are composed of six companies each company consisting of approximately 120 men. Though the government likes to keep the force

[3] Kasturi, Bhashyam, National Security Guards: Past, Present and Future, Bharat Rakshak Monitor, Volume 5 (5), March, 2003. http://www.bharat-rakshak.com/MONITOR/ISSUES-5/Kasturi.html

[4] Special Frontier Force, Wikipedia, http://en.wikipedia.org/Special-Frontier-Force

under wraps, most details are available on the internet. The Special Groups (SGs) have army personnel on deputation. [5]The SFF was employed during the 1971 Indo-Pak war in Bangladesh giving the impression to the Pakistanis that the Chinese were airdropping to assist them. Their photographs are posted on the internet. During the episode of hijack of Indian Airlines Flight 184[6], a Special Group of SFF had also been put on alert. Subsequently, during the 1999 Kargil Conflict between India and Pakistan, employment of the SFF along with pictures was reported by *India Today*, a premier magazine of India.

Combined strength of above Indian Special Forces is perhaps approximate 20,000, more than the than the uniformed strength of US Special Forces (currently around 15,000 that includes Psychological Operations and Civil Affairs Teams who are not uniformed personnel) but not one tenth their capabilities. Additionally, there are efforts to rename Parachute Battalions to "Special Forces (Airborne)" Battalions. We have three Parachute Battalions and two more are under raising. Special Forces must provide focused strategic advantage to the nation but we are incurring tremendous cost on our Special Forces for mere tactical advantages. At current prices, eight Army Special forces battalions are at initial cost of Rupees 2860 crores plus an annual recurring cost Rupees 304 crores. The additional two Army Special Forces battalions will entail an initial cost of Rupees 715 crores and an annual recurring cost of Rupees 76 crores. On balance, 10 Army Special Forces battalions implies initial cost of Rupees 3575 crores plus annual recurring cost of Rupees 380 crores (total Rupees 3955 crores) at current costs. On a very conservative estimate, the approx 20,000 strong Indian Special Forces may cost one and a half times more plus full fledged training facilities: totaling initial cost of approximately Rupees 10,000 crores and yearly recurring cost of approximately Rupees 1,000 crores. If five Parachute Battalions are also considered Special Forces, then there is additional initial cost of Rupees 465 crores and annual recurring cost of Rupees 90.32 crores. These costs are current and will escalate

[5] Photos Special Frontier Force, http://www.militaryphotos.net/forums/showthread.php?62282-very-rare-pics-of-some-of-india-s-Special-Forces-speciality-units

[6] Jaggia, Anil K and Shukla, Saurabh, IC-814 Hijacked: The Inside Story, Roli Books, New Delhi, 2000, pp 59-60.

every year.A much smaller Special Force will cost much less and give tremendous strategic advantage in furthering national security objectives through politico-military operations.

Trans-border employment of Indian Special Forces has been negligible. A serving Special Forces Colonel who has finished command of a Special Forces battalion says, "The Army cannot graduate to the next level of evolution till such time the concept of Special Forces comes to force. But so far it remains just that - a concept. It is unlikely to come about in force till such time the Services shed their 'turf protective' attitude and allow a joint command and control structure. In reference to Army in specific, within the service itself, we have not yet been able to integrate Special Forces with the Aviation or Military Intelligence – thanks to the turf wars between various directorates". Another Special Forces Commanding Officer says, "Apart from very few instances when tasked for trans-border operations, actual employment at present is limited to those infantry tasks that are beyond them due to lack of training, mindset and operational capability. Raids and 'Search and Destroy' are the two main ways the Special Forces are actually tasked. If the Special Forces carry out special operations beyond these, it is usually on their own initiative without the blessings of the command structure."

The weaknesses as observed in Indian Special Forces can be summarised as follows: they are employed tactically; there is no centralised command structure – only country in world, though should be rectified with establishment of a Special Forces Command; no commonality in equipment; no concept of 'packaged equipping' – comes piecemeal; lack of quality manpower and officers – poverty (shortage of officers) shared with the rest of the army; inadequate special / advanced training facilities; lack of integrated, institutionalised and adequate intelligence; grossly inadequate language proficiency; lack of dedicated and integrated air support.; no concept of integral 'support units' including civilian components; army Special Forces more on 'jack of all trades' not specialisation; MARCOS with limited prowess on land; GARUDs sans specialised air transportation units and duplicating tasks of Army Special Forces to large extent; SAGs of NSG and SGs of SFF handicapped with 33 percent annual turnover of manpower; little tri-Services training and nil with NSG and SFF; disparate systems are being acquired for battlefield transparency and network centric conflict and as such, interoperability will remain problematic till such time the issue is addressed.

The USSF have developed a concept of using dogs in the most efficient manner; as buddies of Special Forces operatives that undertake airborne jumps in buddy pairs or even individually and track down their targets in highly professional fashion. None of our Special Forces have made any effort to develop such capabilities aside from routine use of sniffer / tracker dogs as done by the infantry and RR units in counter insurgency areas.

In addition to major voids of institutionalised intelligence and its real time dissemination, and language proficiency is the void of state-of-the-art situational awareness. [7]The US Special Forces have the tremendous advantage in the Special Operations Command Research and Threat Evaluation System (SOCRATES), together with both man portable vehicle borne modules that permit globally deployed Special Forces detachments to receive, send, process and analyse near real time intelligence. Development of a Special Operations Command Post (SOCP) for the PARA (SF) was approved as part of the modernisation of Special Forces in 2001 but has not been developed yet for various reasons, even mundane ones like of single vendor and other reasons of perpetually shifting responsibilities though the vehicle for it was modified years back. The SOCP, once developed, will provide connectivity to Special Forces operatives deployed in field to all required echelons including of aerial weapon platforms.

There is an urgent requirement for creating macro conditions for Special Forces in India. Special Forces of Western nations (and Israel) have conducted many spectacular Special Forces raids, latest being raid to eliminate Osama bin Laden. There have been many snatch missions and dramatic counter-hijack operations like in Entebbe, Mogadishu and London's Iranian Embassy siege. In Indian military literature and writing, we have seldom appreciated hard work, meticulous planning and high level of sophisticated coordination and synergy necessary between various political, military, intelligence agencies and other departments to pull off such missions. There

---

[7] Oberoi, Vijay, Special Forces: Doctrine, Tasking, Equipping and Employment, Centre for Land Warfare Studies, New Delhi, 2006.

is urgent need to educate and create macro conditions for Special Forces through measures like creating a national vision, joint doctrine, joint organisations and integrated intelligence support. Admiral William J Crowe, Jr, former Chairman JCS, USA has this to say, "First break down the wall that has more or less come between Special Forces and the other parts of the military......second, educate the best of the military; spread a recognition and an understanding what you do, why you do it, and how important it is that you do it. Last integrate your effects into the full spectrum of our military capacity." Hopefully, these macro conditions will be created in India concurrent to establishment of a Special Forces Command.

Section III
The Future

1

21st Century Challenges and Warfare 2030

By year 2030, India will be the most populous country and the third largest economy albeit it is difficult to predict what will be the growth rate. Energy and resource scarcity may impede progress unless we optimise resources and develop renewable energy. Our proficiency in managing social change is likely to remain unpredictable. While we will be one of the world's most connected and IT-savvy societies, superpower status depends on where we start out from and how smart are we in managing the evolutionary process. A recent study[1] by London School of Economics says, "India's democracy may have thrived in a manner that few ever expected, but its institutions face profound challenges from embedded nepotism and corruption".

By 2030, our population is likely to be over 1.53 billion at current annual growth rate of 1.58 percent. 65 percent of current population is 35 years and below and we have the largest illiterate population in the world. Then is the problem of unemployment, which stands at 10.5 percent up to the 35 year age group.[2] Some 40 million illegal weapons are currently circulating within India with an annual trade of $4 billion. [3] The World Drug Report

---

[1] India: The Next Super Power, *The Economic Times*, March 7, 2012 http://articleseconomictimes.indiatimes.com/2012-03-07/news/31132262_1_superpower-state-hillary-clinton-Ise

[2] Overdorf, James and Teng, Poh,India: Illegal guns plague cities, Globspot December 20, 2010.http://www.globspot.com/dispatch/India/101214/India-illegal-guns-gun-control-crime

[3] World Drug Report 2012, http://www.undoc.org/undoc/en/data-and-analysis/WDR-2012.html

2011 shows some 3.2 metric tons of drugs entered India through Pakistan in 2009 (one single year). [4]We are 134th in Human Development Index 178 countries since past decade and [5]95th in corruption perception index amongst 183 countries as per Transparency International. This is a dangerous cocktail for terrorist groups luring youth even for mere fiscal benefit.

In addition to causes of conflict like territory and power, resources like water, energy, minerals etc will be the major flash points. This will see a heightened need for intelligence and deniable covert capabilities that ensure deniability of action for achieving strategic aims, both of which will have a major Special Forces component. There is little doubt that asymmetric wars (of which terrorism and insurgencies are manifestations) will continue to dominate the conflict spectrum in the Sub Continent albeit windows of conventional war under the NBC backdrop will remain. Warfare is no longer confined to the battlefield - boundaries between war and no war are blurred by asymmetric wars that have no borders, no rules and no regulations. Psychological warfare probably imposes the largest penalty but affords the highest payoffs. Successful psychological warfare demands integrated themes and subjects which need to be developed.

21st Century conflicts will have the following connotations: ability to engage in armed conflict in nuclear backdrop will remain an instrument of state power; sub-conventional conflicts characterised by intra-state strife have gained ascendency over traditional conflicts; spectrum of conflict could range from conflicts between states to conflict with non-state actors and proxies; transnational nature of threats and involvement of state actors in using sub-conventional conflicts have increased the complexity; conventional conflict could either be preceded, in conjunction or succeeded by a period of irregular conflict, requiring LIC and stabilisation operations; non-state actors have added a new dimension to Low Intensity Conflict (LIC) and they are increasingly acquiring conventional capabilities; wars will be short, swift, hi-tech invariably fought in complex terrain, requiring full spectrum

---

[4] India Ranks 134 in human development index,*Hindustan Times*, September 30, 2012. http://www.hindustantimes.com/News-Feed/India-ranks-134-in-human-development-index/Article1-76401.aspx

[5] India 95th among 183 countries in Corruption Perception Index, *The Economic Times*, December 1, 2011. http://articles.economictimes.indiatimes//2011-12-01/news/30412959_1_corruption-perception-index-ranks-countries-cpi

capability; technology empowers terrorists to cause severe damage through cyber, financial, kinetic attacks – their acquiring WMDs is a major concern; WMDs will proliferate exponentially; proxy war will remain preferred strategy of weaker / radicalised nations; scarce natural resources could lead to conflict; friction between state militaries and non-state actors is likely to increase.

What should be of military security concern to India, can be summarised as follows: collusive China-Pakistan threat and conflict within India (two and a half front); possibility of US-Pakistan mutual dependency remaining critical viz-a-viz US-India needs in strategic and security values even beyond 2014; greater Pakistan Taliban control of AfPak post 2014, namesake democracy in Pakistan and increasing radicalisation; Chinese presence in Pakistan / Pakistan Occupied Kashmir and China's tacit support to Pakistan's jihadi policy against India; Pakistan's future proxy war cashing upon carefully nurtured home-grown radicals in India; hardening of Chinese stance on Arunachal and IOR with increasing military capability viz-a-viz India; increased Chinese dominance in waters of Asia-Pacific; Chinese 'String of Pearls' surrounding India plus the 'Ring of Islamic Terrorism'; China engineered Maoists in Nepal and their integration with Nepal's Military; change of guard in Bangladesh with Khaleda Zia and her BNP regaining power can see revival of terrorism; presence of PLA and Chinese Special Forces in garb of development projects in countries surrounding India; radicalisation of Maldives and possibility of LeT bases coming up in over 1000 uninhabited islands in that country; LeT and Nepalese Maoists links with Indian Maoists; growing interdependence and interlinking of terrorist groups and insurgents pan India and their international links, particularly Islamic radicals like Indian Mujahideen (IM) and Popular Front of India (PFI) links with Al qaeda and LeT; management of social change within India, without which, youth may polarise towards terrorism / insurgency; heightened asymmetric war including cyber war, plus activation of space, cyberspace and electromagnetic domains.

Huma Yusuf wrote in *The Dawn* in October 2011[6], "In the absence of the activism of democracy, you are left with the fatalism of patronage. A

---

[6] Yusuf, Huma, Quest for a Dengue Free Democracy, *The Dawn*, October 7, 2011.

nation that obsesses over external threats is one that values patronage, because patronage means protection from what may come. Valuing patronage is in some ways the antithesis of voting in a democracy: rather than shape your future, you seek protection from it. Ironically, patronage also nullifies the future possibility of democracy because it reiterates the importance of that which is local — kinship, ethnicity, language, sect — over what is national. As long as we seek protection from an external enemy, we will seek patrons, even if they come in uniform —and it is thus that history readies to repeat itself."

More significantly, [7]Pervez Hoodbhoy, Professor of Nuclear and High Energy Physics, Quaid-e-Azam University, Islamabad stated on 18 Aug 2011, "An extremist takeover of Pakistan is probably no further than five to 10 years away".[8]Pervez Hoodbhoy also warns of the institutionalised radicalisation in Pakistan by saying, "The common belief in Pakistan is that Islamic radicalism is a problem only in FATA, and that madrassas are the only institutions serving as jihad factories. This is a serious misconception. Extremism is breeding at a ferocious rate in public and private schools within Pakistan's towns and cities. Left unchallenged, this education will produce a generation incapable of co-existing with anyone except strictly their own kind. The mindset it creates may eventually lead to Pakistan's demise as a nation state."

A 2009 study[9] titled 'Beyond Bullets: Strategies for Countering Violent Terrorism' by the Centre for A New American Security, USA states, "Some countries require particular attention. Pakistan, for example, represents the most difficult problem because it has become the host of the global jihadist movement and terrorists can increasingly operate with impunity there because of the weakening of the state". The fact is that the linkages of Pakistan's

---

[7] Hoodbhoy, Pervez,An Extremist Takeover of Pakistan is Probably No Further than Five to 10 Years Away, *The Dawn*, February 26, 2011. http://www.3quarksdaily.com/3quarksdaily/2011/02/an-extremist-takeover-of-pakistan-is-probably-no-further-than-five-to-10-years-away.html

[8] Hoodbhoy, Pervez, The Suadi-isation of Pakistan,http://groups.yahoo.com/group/Writers_Forum/message/39644

[9] Beyond Bullets: Strategies for Countering Violent Terrorism, Study by the Centre for A New American Security http://www.cnas.org/files/documents/publications/LordNagRosen-Beyond%20Bullets%20edited%20Volume-june09_0.pdf

Military-ISI with terrorist organisations in Afghanistan, Pakistan, India and Bangladesh have serious implications for us. Pakistan's refusal to act against the Haqqanis is proof of her intentions in Afghanistan post 2014.

It is no secret that Pakistan has adopted terrorism as a state policy. During the Regional Conference on Security held in Bangladesh in 2001, both Pakistani speakers (Dr Shirin Mazari, DG Institute of Strategic Studies and Lt Gen Javed Hassan, Commandant, National Defense College) propagated LIC and unconventional means like guerilla warfare, psychological warfare, including the use of terror, subversion, economic warfare and indirect intervention in enemy territory as more viable options to a conventional war. A decade later, Admiral Mike Mullen confirmed this as the State Policy of Pakistan before the US Senate Armed Services Committee on Afghanistan and Iraq. Pakistani apparently adopted this policy on advice given by [10]Chinese Premier to President Ayub Khan of Pakistan in 1960's that Pakistan should prepare for prolonged conflict with India instead of short-term wars, advising Pakistan to raise a militia force to act behind enemy lines. Pakistan obviously took the above advice seriously. Not only was cross border terrorism initiated, a terrorist architecture spanning India was orchestrated. [11]MK Dhar, Former Joint Director IB writes in his book 'Open Secrets–India's Intelligence Unveiled', "Way back in 1992-93 ….the process of 'transplanting armed modules' in the heartland of India had started taking cognizable shape. Some of these cells were identified in Assam, West Bengal, Bihar, Uttar Pradesh, Delhi, Kota/Ajmer region of Rajasthan, Gujarat, Maharashtra, Andhra Pradesh, and Kerala. SIMI had already started deputing 'volunteers' to Pakistan for training along with the Mujahideen, Taliban and Al Qaeda cadres. They established firm linkages with Islamic Chhatra Shibir, Al Qaeda affiliated HUJI, Al Badr, Al Jihad and other organisations in Bangladesh ……. were trained in facilities located inside Bangladesh and under the very noses of DGFI and BDR." Ajmal Qasab admitted that the 26/11 Mumbai attack terrorists were trained by Pakistani Marines, intelligence has established that Indian Mujahedeen (IM) are the creation of Pakistan's Inter Services

---

[10] Aijazuddin, FS, From a Head, Through a Head, To a Head – The Secret Channel between the US and China through Pakistan, Oxford University Press, Karachi, 2000.

[11] Dhar, Krishna, Maloy, Open Secrets – India's Intelligence Unveiled, Manas Publications, New Delhi, 2005.

Intelligence (ISI) and much more damaging evidence has come from Abu Jundal, all of which Pakistan blatantly denies. Presently, the Lashkar-e-Taiba (LeT) is engaged in indoctrination, training and arming Maldivian youth frequenting Pakistan.

There should be little doubt that the ISI and through them the Pakistani Military are linked with Al Qaeda, Pakistan Taliban, Haqqani network, LeT, HUJI, HuM, LeJ, Dawood Ibrahim, in Bangladesh with AHAB, Huji BD, JMB, JMJB, and in India with SIMI, IM, Maoists, Popular Front of India either directly or through LeT which serves as the covert arm of ISI. Over the past decade, the LeT which was originally raised to undertake jihad against India has become more lethal internationally; more powerful if not more than Al Qaeda, which led [12]Ashley J. Tellis (Carnegie Foundation) to state in March 2012, "The only reasonable objective for the United States is the permanent evisceration of LeT and other vicious South Asian terrorist groups—with Pakistani cooperation if possible, but without it if necessary." [13]Bruce Riedel (Brookings Institution) added in July 2012, "With Al Qaeda on the ropes, Lashkar-e-Taiba (LeT), with the help of its Pakistani backers, is now probably the most dangerous terror group in the world." [14]As for Taliban, Murtaza Ali Shah wrote in the Dawn dated 27 Oct 2011, "Taliban commanders acknowledge in BBC documentary titled 'Secret Pakistan' that Pakistan is still running training camps to support and arm the Taliban across the border in Afghanistan."

How the continuing China-Pakistan nexus is going to affect Afghanistan and the prospects of Indo-Pak peace has been very lucidly brought out by [15]Aga H Amin, defence analyst and veteran Pakistan Army officer, wherein he says, "Utopians in India are jubilant that Pakistan has made peace with India. Nothing in reality can be farther from the truth. ….. Pakistan's

---

[12]Tellis, J, Ashley, The Menace That Is Lashkar-e-Taiba, *Policy Outlook*, March 2012, http://carnegieendowment.org/2012/03/13/menace-that-is-lashkar-e-taiba

[13] Riedel, Bruce, Mumbai terror Attack Group Lashkar-e-Tayyiba Now More dangerous Than Al Qaeda, http://www.dailybeast.com/articles/2012/07/02/mumbai-terror-attack-group-lashkar-e-taiyyaba-noe-more- dangerous-than-al-qaeda. html/

[14] Shah, Ali, Murtaza,Secret Pakistan, *Dawn*, October 27, 2011.

[15] Amin, H, Agha, Can India and Pakistan Make Peace,http://reportmysignalpm.blogspot.in/2012/08/india-and-pakistan-make-peace-by-major.html

apparent shift is merely a tactical response to extreme confrontation with the US over perceived US view that Pakistan is playing a double game in Afghanistan.... The real picture of true intentions of the Pakistani military will emerge when the US withdraws from Afghanistan. This will be the time when the Russians, Iranians and Indians will have no choice but to support the Northern Alliance against Pakistan sponsored Taliban who regard all Shias, Ismailis, Non-Pashtuns, moderate Pashtuns as infidels who deserve to be massacred. .... Pakistani politicians will remain the puppets of the military that they have been since 1977. Terrorism will remain a tool of foreign policy while the Pakistani military runs the Pakistani state under a facade of PPP or PML or Tehrik i Insaaf......Pakistani military will be hoping to achieve all its objectives: an extremist dominated Afghanistan; a Baluchistan fully fragmented and crushed; a Pakistani political party leading Pakistan fully subservient to the Pakistani military; a renewed infiltration in Kashmir; a brinkman's nuclear policy with India; and a greater Chinese vassal with far greater Chinese interests in Pakistan. There is no doubt that Pakistan will be a semi autonomous Chinese province by 2030 or so. .... Pakistani Baluchistan by 2030 would be a completely Chinese run show... This means that Pakistan's political economy of exporting terrorism as a foreign policy tool, massive corruption at home and the resultant ever growing reservoir of economically deprived youngsters who will fill ranks of extremists and suicide bombers will continue".

The emergence of irregular / asymmetric forces with greater strategic value over conventional and even irregular forces in conflict situations over conventional and even nuclear forces in recent years has not been acknowledged by India. Consequently, we have failed to create deterrence against irregular forces relying only on diplomacy, which itself is not fully effective not having been sensitised to military and particularly to the advantages of Special Forces. For instance, after the Anglo-Franco-Israeli assault on Egypt in 1956, the British had been booted out of the Middle East. A few years later the British Foreign Office offered the services of the SAS as 'advisers' and experts to help the middle-east regimes quell their insurgencies and meet other security needs. This led to the British regaining their influence in the region and re-emerge as a major foreign affairs player in the energy-sensitive region. Similar initiatives have been undertaken by the US and Israeli Special Forces. We continue to remain at great disadvantage. Lack of strategic forethought has deterred us to exploit the strategic potential of our Special Forces in this regard. Amy Kazmin wrote in *Financial Times* dated 9th Sep 2011, saying, "The instruments of State action have become dysfunctional", says K Shankar Bajpai, Chairman of India's National Security Advisory Board and former Ambassador to US. India's strategic interests extend between the Suez to Shanghai .... but we have neither the manpower nor the strategic thinking to handle these challenges". There is urgent need to develop publicised overt capabilities and deniable covert capabilities as deterrence against irregular war thrust upon us plus the will to selectively demonstrate it to ensure its effectiveness. Our Special Forces potential must be optimised and exploited to develop such capability.

2

Foreign Special Forces Employment

Special Forces employment over the years can be gauged from the following: German pseudo teams in Ardennes during in WW II; British pseudo teams against communist guerillas and in Kenya against the Mau–Mau insurgents; Selous Scouts raised in Rhodesia (1973) were responsible for 68 percent of all terrorists killed in the next six years of Bush War at cost of 40 Scouts; USSF employed in Afghanistan one and a half year prior to US invasion; Special Forces of NATO countries operating in Afghanistan, Syria, some African countries; Russia's Spetznaz deployed in CIS, Eastern bloc countries and even in Pakistan; Pakistan's SSG has been operating in J&K, Afghanistan, Iraq, Nepal, and Bangladesh; USSF in Iran since past seven years, operating in some 200 countries (including 186 sanctioned in 2011). Afghanistan provided a challenge for Special Forces operations. In addition to Special Forces of many NATO countries, deployment of Green Berets, Delta Force and Navy Seals from the US were deployed at various points of time, and so were the SAS and SBS of Britain. A plethora of Special Forces operation in close proximity in the same region as being done by us in J&K, has its own attenuated problems; when there is lack of elbow space and everyone is looking for awards and glory. In the case of Israeli Special Forces, their erstwhile role of mainly long range reconnaissance, intelligence and other infantry like tasks has changed focus over the years to counter terrorism and hostage rescue. Most of these units, with training on advanced counter terrorism, have been employed successfully in operations.

Right from its raising, Pakistan's SSG has had close training and advisory links with US Special Forces. In fact, Pakistani officers had been visiting US SOCOM from the time the latter was being set up. The SSG's first major operation, however, was during the 1965 India-Pakistan War wherein the SSG were dropped in the vicinity of IAF airbases of Adampur, Halwara and Pathankot. This turned out a complete fiasco; most SSG personnel were taken prisoners without damaging a single combat aircraft. Again during 1971 India-Pakistan War, SSG was used extensively employed in what now is Bangladesh. One particular mission in raiding a gun position was successful.

Pakistan denies the use of SSG during the 1999 Kargil Conflict with India but SSG were used extensively mixed with the Northern Light Infantry troops, with Pakistan showcasing the intruders as freedom fighters till the lie was exposed to the world. The SSG must have taken considerable casualties particularly with heavy artillery pounding though the figure will remain unknown. During the same period, an SSG detachment attempted capture of the glaciated peak of Point 5770 in the Southern Glacier on the Saltoro Range in Siachen area. However, they were annihilated through a daring daytime raid by the Indians. The bodies of dead Pakistanis including that of an SSG Captain were later handed over to Pakistan on their behest.

Where Pakistan's SSG gathered considerable operational experience was in Afghanistan fighting alongside the Taliban against the Russians. SSG have also been employed in Kashmir and for training terrorist in Nepal and Bangladesh at various points of times. SSG has large presence in many Arab countries in training and advisory capacities focused on counter insurgency and VIP protection. They have a pivotal role in protecting the Saudi royal family. They have extensively trained with USSF, Sri Lankan Special Forces and are now increasingly undertaking joint training with Chinese Special Forces.

Chinese Special Forces numbering an estimated 14,000 are specialised in rapid reaction combat in a limited regional war under high-tech conditions, commando operations, counter-terrorism, and intelligence gathering. They are organised in a number of units and sub-units like: Guangzhou Military Region Special Forces Unit established in 1988 and expanded in 2000 as first PLA special operations unit capable of air, sea and land operations; Chengdu Military Region Special Forces Unit established in 1992 and specialised in target surveillance, target designation, airborne insertion, sabotage, offensive strike, rescue, and has experimented new concepts, tactics, equipment including digitised army soldier system and high-mobility land weapon platforms; Beijing Military Region Special Forces Unit established in early 1990s and holds high-tech equipment including UAVs, modern demolitions, laser designators and laser dazzlers; Shenyang Military Region Special Forces Unit; Nanjing Military Region Special Forces Units (two of them), Nanjing Military Region Special Forces Unit; Lanzhou Military Region Special Forces Unit; Hong Kong Special Operations Company known as five minute Response Unit; and the Macau Quick Reaction Platoon. In conflict scenarios, Chinese Special Forces will likely be deployed in conjunction with China's Airborne Corps. However, in non-war period Chinese Special Forces would be covertly deployed for information support operations, strategic surveillance, training, arming and advising dissident / terrorist / insurgent groups in target countries, and perception management.

If China aims to stir up NSCN, ULFA and Maoists insurgencies while claiming Arunachal Pradesh and illegally occupying Aksai Chin and Shaksgam Valley, can we rule out covert involvement, including by proxy, of their intelligence agencies and Special Forces? Covert presence of Chinese Special Forces and intelligence agencies in Pakistan, POK, Nepal, Myanmar, Bangladesh, Sri Lanka and India is real possibility with PLA owned construction companies spread over globally under garb of development projects. In India, they could have presence in the Chinese Embassy, business ventures and border areas where facial features don't pose any problems and illegal smuggling routes provide avenues for ingress and contacts. Some Chinese nationals with fake Indian documents have already been apprehended in a bid to contact northeast insurgents. What needs to be remembered is that Mao's guerilla war had been ingrained in the PLA for decades to start with, followed by Chinese assumptions that future military contingencies could come up without much warning and may involve smaller forces but with highest operational potential and readiness. If this is coupled with the Chinese strategy of surprise, shock and keeping the adversary lulled, one should logically surmise that akin to cyber capabilities, the potential of Chinese Special Forces would actually be much more than what they would like the world to believe.

The world, India in particular, should watch out for congruence of Chinese and Pakistani Special Forces interlaced with terrorist organisations like the Taliban (with whom both China and Pakistan are already linked), Al Qaeda, Haqqani network, LeT, Hum etc. This is more relevant with expanding Chinese strategic footprints in Pakistan Occupied Kashmir (POK) and likely leasing out of Gilgit-Baltistan area by Pakistan to China for the next 50 years as reported in Pakistani vernacular media plus US media and Think Tanks. To top this, Pakistani defence analysts predict Pakistan likely to go in for a similar arrangement in Baluchistan post 2014. Enhanced training of Chinese and Pakistani Special Forces in seaborne and underwater training indicates countries in the Indian Ocean Region can be targeted along with threat to coastal regions. Uninhabited islands could be used for establishing bases as springboards for Special Forces / terrorist operations.

In recent times, the West has been employing its Special Forces for regime change as a new asymmetric option / policy. The combination is Information Warfare (IW), Intelligence agencies, Special Forces and air power applied in the last stages. If media reports are to be believed, US engages in asymmetric war by any means in national interest; recent reports indicate US used Al Qaeda in Libya and is doing similarly in Syria in conjunction USSF, NATO, Turkish and Qatari Special Forces mixed with rebels / opposition.[1] Paul Joseph Watson, wrote in December last that just as Al-Qaeda terrorists were used to oust Gaddafi, hundreds of Libyan rebels with Al Qaeda willing members were being airlifted into Syria to aid opposition in carrying out attacks against government forces. This implies USSF is using willing captured Al Qaeda cadres including detainees from Guantanamo prison.[2] This is a clever move and no different from World War II where the US OSS (Office of Strategic Services), predecessor to USSF, utilized members of a German dissident group who had fled to France as refugees for unconventional operations against the German Army in conjunction OSS.

---

[1] Watson, Joseph, Paul, US Special Forces Mass on Syrian Border, infowars.com, 13 December, 2011, http://www.infowars.com/us-special-forces-mass-on-syrian-border/

[2] Bank, Aaron, From OSS to Green Beret – The Birth of Special Forces, Presidio Publishers, USA, 1986.

The philosophy of the then OSS Chief, General Donovan had been, "Use them as long as they kill Nazis".[3] In October 2012, Mitt Romney, US Presidential candidate vowed to arm Al Qaeda in Syria, responding to which Paul Joseph Watson, Editor of PrisonPlanet wondered whether America shares its values with terrorist.

Core tasks of foreign Special Forces are direct action, special reconnaissance, foreign internal defence, unconventional warfare, counterterrorism, and counter proliferation of WMDs, civil affairs operations, psychological operations and information operations. It is significant to note that the USSF also undertakes undeclared tasks like conducting proactive, sustained 'man hunting' and disruption operations globally, building partner capacity in relevant ground, air and maritime capabilities in scores of countries on a steady - state basis., helping generate persistent ground, air and maritime surveillance and strike coverage over "under-governed" areas and littoral zones, and employing unconventional warfare against state sponsored terrorism and trans-national terrorist groups globally. Additionally, the CIA had a 3000 strong army of Counter Terrorism Pursuit Teams (CTPT) in AfPak region that have depleted in Pakistan post the Raymond Davies episode and killing of Osama-bin-Laden.

---

[3] Watson, Joseph, Paul, Whether America Shares Its Values with Terrorists, prisonplanet.com, http://www.prisonplanet.com/

3

Why India Needs Special Forces

India has been suffering glaring intelligence voids continuously. We became aware of Chinese building a highway in Aksai Chin 11 months after construction began and just one month before its completion. Ouster of 3,500 Indian business families from Fiji and their replacement by Chinese came as a surprise. Chinese activities in Coco Island intrigued us for many years. We had no prior information of Bhutanese Consulate established in Hong Kong, Kargil intrusions, the royal massacre and Maoist build up in Nepal, and two attempted coups in Bangladesh. More recently, we had no inkling of some 11,000 Chinese having forayed into Pak / POK, the trouble in Maldives and the MNDF (Maldives National Defence Force) leanings towards a possible Maldives-China defence pact. There is little that we know of PLA disguised in development projects abroad. We do not have requisite arrangement for surveillance in areas of our strategic interest. If we had Special Forces operatives deployed in areas of our strategic interest, we would not be blind to future threats. The US discovered technical intelligence is not enough when Pokhran II fooled and surprised them completely. That is the reason they have Special Forces operatives inside Iran for past several years. Had India deployed Special Forces cross border on politico-military missions, we would have been better prepared to deal with the following: LTTE turnabout – our Special Forces would have infiltrated them, their intentions known and their leadership targetable in event of double cross; situation in Tibet, Xinjiang and Pakistan-China occupied Kashmir for using these to India's advantage; assassination of Sheikh Mujibur Rehman and crashing of the Indian helicopter meant to evacuate him – advance information could have avoided these incidents; Chinese support to Maoist in Nepal and Chinese activities in Nepal, Bhutan, Myanmar, Bangladesh, Sri Lanka, Afghanistan, CAR, Seychelles, and IOR etc; Chinese links and support to ULFA, weapon support to Indian Maoists and Kachin rebels in north Myanmar and through latter to PLA in Manipur and Indian Maoists; IC-814 hijack, Parliament attack, 26/11 and other Pakistan sponsored terrorist attacks. It is significant to note that a 26/11 type seaborne attack was originally planned by Al Qaeda against Manhattan in New York but US SF could infiltrate Al Qaeda and thwarted it because of advance information. The state sponsored anti-India forces too need to be infiltrated and targeted within Pakistan.

Most of the above scenarios for covert deployment of Special Forces remain relevant. Though it will not be prudent to mention specific future requirements due to security constraints, following can be pointers:Chinese exercises / activities in close proximity of the border can raise alarm, as in present case of R&AW suggesting skirmish / conflict, unless we have Special Forces operatives on ground monitoring movements and assessing the situation continuously; considering the fluid situation, there has been speculation of balkanization of Nepal which may be orchestrated by China through the Maoists with possibility of certain areas being usurped by China. The situation requires close monitoring that cannot be done through technical means; Chinese forays into Bhutan and India need as much advance warning as possible considering the state of our forward communications that inhibit timely mobilization and are unlikely to improve in near future. Only Special Forces can provide such advance warning and impose delays on the enemy; the "grab for grab" policy advocated by Nonalignment 2.0 can hardly be executed without use of the third dimension, which is not possible unless you already have Special Forces operatives on ground; in the event of an out of area contingency as in the past in Maldives, airfields may not be available if held by terrorists / rebels unless we had Special Forces already on ground for advance information and facilitation; Chinese-Pakistani assistance to aggravate turmoil in India by supporting, arming and unifying insurgents in India cannot be countered without deploying Special Forces strategically on politico-military missions to create necessary deterrence.

We must be prepared for manifold heating up of the internal security situation by 2030. Our population by then would have crossed 1.53 billion at current annual growth rate of 1.58 percent. For past decade, we are content to be constant at 134th in human development index and 95th in corruption perception index. By admission of the Planning Commission itself only 17 percent of money spent on poverty alleviation schemes reaches ground level. 65 percent of our current population is 35 years and below. We have the largest illiterate population in the world and current unemployment rate is 10.5 percent. As per estimates, 40 million illegal weapons are already circulating within India with an annual trade of $4 billion. To top this, World Drug Report 2011 talks of some 3.2 metric tons of drugs having entered India through Pakistan in one single year (2009). Our inability to manage social change therefore will be exploited by both China and Pakistan. India needs to organise its Special Forces to deal with this turmoil.

India cannot afford to be lulled into complacency and visions of regional peace when both China and Pakistan are continuously waging asymmetric war on us; cross border terrorism, support to indigenous terrorist organisations (including weapons, communication equipment, financing, training and indoctrination), drugs, fake currency, propaganda, inflaming communal violence and the like. The ISI surrogate terrorist organisations pose extreme danger. It is this understanding that led [1]Ashley J. Tellis of Carnegie Foundation to state in March 2012, "The only reasonable objective for the United States is the permanent evisceration of LeT and other vicious South Asian terrorist groups—with Pakistani cooperation if possible, but without it if necessary."

For the same reason, MK Dhar[2], former Joint Director Intelligence Bureau wrote, "I continued to advocate for an aggressive and proactive counter and forward intelligence thrust against Pakistan. My voice was rarely heard and mostly ignored. The Pakistani establishment is a geopolitical bully. The best response to blunt such a bully is to take the war inside his home. India has allowed itself to be blackmailed by Pakistan even before it

---

[1] Tellis, Ashley, The Menace That Is Lashkar-e-Taiba, Carnegie Foundation for International Peace, Policy Outlook, March 2012, http://carnegieendowment.org/2012/03/13/menace-that-is-lashkar-e-taiba

[2] Dhar, Krishna, Maloy, Top Secret – India's Intelligence Unveiled, Manas Publications, New Delhi, 2005.

went nuclear. The sabre rattling of coercive diplomacy, which is nothing but sterile military power, cannot convince the Islamist Pakistani Establishment that India can take the border skirmishes inside their homes and hit at the very roots of the jaundiced islamist groups." The Pakistani Military-ISI maintain their links with the Taliban, despite spats. This is very apparent with statements from Taliban leaders like Maulvi Fazlullah.

Chinese links to Taliban go way before the US invasion of Afghanistan in 2001. Post 2014, once Pakistan achieves her strategic depth in Afghanistan and China consolidates in Pakistan Occupied Kashmir, there is no telling if the Taliban will also be unleashed on India. This should also be reason enough for India to think hard before agreeing to withdraw from Siachen, since in doing so we will be opening the floodgates of infiltration into Ladakh where ISI has already attempted fomenting communal tensions between Buddhists and Muslims. ISI has been nurturing Shia terrorist outfits with an eye on Ladakh since late 1990s. When Musharraf said, "There will be many more Kargil's", it was with the confidence that he will be able to fool India into withdrawing from Siachen.

If we remain ambivalent and do not employ our Special Forces pro-actively, China and Pakistan will be able to unify terrorists and insurgent outfits in India and create conditions of near civil war in the hinterland particularly using the Maoists. The two and a half front presently being talked about can well become a three front threat. Ambivalence amounts to cowardice in such scenario. [3]Even Gandhi Ji, Father of The Nation and the apostle of non-violence spoke thus to Brig (Later Lt Gen) LP Sen, Commander (Designate) 191 Infantry Brigade in 1947, "If one has to choose between cowardice and violence, I will choose violence". *The Asian Age* dated 11 March 2012 had quoted Afghan scholars visiting New Delhi in saying, "Shed 'Gandhigiri' approach with Pakistan". We should heed Chanakya, who had said "Do not be very upright in your dealings for you would see by going to the forest that straight trees are cut down while crooked ones are left standing".

---

[3] Sen, Lionel, Pratip, Slender Was the Thread : Kashmir Confrontation 1947-1948,Orient Longman Ltd, New Delhi, 1969.

As mentioned earlier, the emergence of irregular / asymmetric forces with greater strategic value over conventional and even irregular forces in conflict situations over conventional and even nuclear forces in recent years has not been acknowledged by India. Consequently, we have failed to create deterrence against irregular forces relying only on diplomacy, which itself is not fully effective not having been sensitised to military and particularly to the advantages of Special Forces. Special Forces form an essential part of military diplomacy in molding perceptions, in line with the national security objectives of the country and act as an extension of foreign policy. We therefore, will continue to remain at great disadvantage at the mercy of China, Pakistan and their proxies unless we optimise our Special Forces potential and create necessary deterrence against irregular war thrust upon us.

4

Doctrinal and Conceptual Issues

The debate on Special Forces in India hitherto has been caught in the maize of Special Forces, Commando Forces and Airborne Forces and their employment within India that naturally overlap their tasks. By and large strategic tasking has not crossed the realm of the theoretical. Globally, Special Forces have always been synonymous with national strategic interests. What has happened in India is that in absence of a National Security Strategy and national security objectives clearly defined, the Special Forces have actually landed up being 'commando troops' although we already have commando platoons in the hundreds of Infantry and Rashtriya Rifles battalions that have really not been optimised as commando troops because the former are available in every theatre. In a whole set of seminars held over the years the need, necessity, imperatives, quantum and control of Special Forces required has actually not been addressed holistically, particularly the quantum and control of not the existing commando type of forces but the actual Special Forces that are meant for Strategic employment. Resultantly, a clear cut policy for employment of Special Forces at the national level has not been evolved, partly because of lack of strategic forethought and partly because the controlling masters of the various Special Forces entities are cozy in respective turfs and cocooned environment.

In October 2008[1], HQ Integrated Defence Staff (IDS) released a Joint Doctrine for Special Forces that was widely covered by Indian media. The Doctrine was in recognition of global acknowledgement that all future wars

---

[1] Joint Doctrine for Special Forces Unveiled, *The Hindu*, 02 October, 2008, http://www.hindu.com/2006/10/2/stories/200810025607130.htm

and conflicts will require the Special Forces to play an increasingly important role at all levels of war; strategic, operational or tactical. The doctrine covered the military Special Forces, the operational environment in which the Special Forces are likely to operate, organisational set-up and special characteristics of the Special Forces and charting out the ideal command and control organisation and joint planning at theatre level. The doctrine emphasized the need for accurate intelligence, aerial fire support, naval gunfire, artillery, precision-guided munitions and rockets for the successful conduct of special operations. It also highlights the importance of various aspects of joint training to achieve greater cohesion and understanding necessary for joint special operations. Since then, there has been quantum increase in joint training with foreign Special Forces but not much follow up on the doctrine per se though Special Forces from the Army and the Navy had been extensively employed in operations in Sri Lanka in the past.

The Army Special Forces in war have the tasks of Reconnaissance, Surveillance and Target Designation (RSTAD) and Special Operations. In counter insurgency and counter terrorism their tasks are Surveillance and Reconnaissance, Covert Operations, Pre-emptive / Retaliatory Trans-Border Operations and Hostage Rescue of armed forces personnel. During peace time they are tasked for Special Missions and Out of Area Contingencies. Special Operations, Covert Operations and Special Missions provide the leeway to undertake virtually any strategic task. However, this has really not taken off for lack of politico-military will.

Notwithstanding the above, major doctrinal change in a country like the US too did not come about overnight. In fact, inter-service integration virtually had to be forced down upon the military because of a series of incidents over the years like the failed Iranian hostage rescue attempt, the terrorist attack on Beirut resulting in large number of American casualties and similarly the problems of interoperability and yet again many casualties suffered during the US operations in Grenada. The operation in Grenada brought home the utter lack of cohesion between the Services. Lessons from these led to the establishment of strategic command and eventually US Special Operations Command (SOCOM). Learning from global development of Special Forces and analyses of operations undertaken by them, it is possible to evolve a doctrine and concept of employment of

Special Forces suited to Indian requirements. We do not really have to go through the same evolution process for organising integrated Special Forces but the urgency for achieving synergy is very much continuing. The fact remains that our Ministry of Defence (MoD) has no uniformed personnel permanently absorbed or posted on deputation, the HQ Integrated Defence Staff (IDS) has come up akin to a separate Service HQ, the Services lack jointness and as such the Military is out of the strategic thought process at national level.

Fault-lines occur for various reasons, however, one of them is lack of governance or faulty-governance that generates dissent and dissatisfaction, which left unaddressed, can turn into individuals / groups taking up arms. This is a shade different from 'perceived injustice' that the faulty or corrupt administration would like to convey to outsiders. A government's inability or unwillingness to meet the legitimate needs of its people may result in popular frustration and dissatisfaction. People may lose their faith and confidence because the government lacks legitimacy. Propaganda apart, this is one of the fundamental problems. You can find numerous examples of this, especially in developing countries.

Dissent again has many connotations, for example, curbs on personal independence and banning social networking in China has resulted in large segments of simmering population. Because of such phobia, China had even banned a simple set of exercises like Falun Gong fearing this would give opportunity to large bodies of population to come together with collective thinking going at tangent to what the CCCP wants the population to adhere to. But the internet has negated such bans with software available that can access blocked sites. It is the fault-lines or rather the dissent that the adversaries cash upon. Special Forces do not create resistance movements but advice, train and assist resistance movements already in existence. They are ideally suited to control fault lines of the adversaries without any signatures or with ambiguous signatures. This is precisely what China and Pakistan are doing as part of asymmetric war on India, using proxies. [2]Understandably, it is no easy task as it has considerable gestation period involving building

---

[2] US Army Field Manual 31 – 20, Doctrine for Special Forces Operations, http://www.enlisted.info/field-mannuals/fm-31-20-doctrine-for-special-forces-operations.shtml

rapport with and adapting to the ways of an indigenous resistance outfit, and eventually take control and direct actions for maximum effect. There is urgent need to optimise our Special Forces potential to correct this asymmetry.

According to Stephen Cohen[3], "The task of Special Forces is the proxy application of force at low and precisely calculated levels, the objective being to achieve some political effect, not a battlefield victory." While Special Forces should be central to asymmetric response including against irregular forces, asymmetric warfare does not automatically equate to a physical attack. A physical attack is only the extreme and potentially most dangerous expression of asymmetric warfare. The key lies in achieving strategic objectives through application of modest resources with the essential psychological element. In asymmetric settings, Special Forces have limitless pro-active employment possibilities to exploit dissidence; employ asymmetric approaches from the NBC sphere to psychological operations, information war, economic / technical / financial war. In counter terrorism and counter insurgency, Special Forces can be used for Intelligence, surveillance and psychological operations, rival / pseudo gang operations, infiltrating *tanzeems,* neutralising terrorist leaders, organisations, support groups, infrastructure, selective raids, ambushes, snatch operations and incident response operations. In out of area contingencies, they can assist airborne / conventional forces or may be called upon to perform politico-military missions like assistance to third world nations, surgical strikes, recovery missions, prevent terrorist use of WMDs, humanitarian assistance etc.

Special Forces have a covert role not only in irregular combat including training dissidents but also through psychological operations, perception management and other noncombat activities including economic, social and cultural, all of which require focused planning including cross-cultural communications, and preparation before the mission is launched. It goes without saying that Special Forces operatives in the field must have access to all time situational awareness, real time / near real time operational picture and regularly fed assessments and forecasts. [4]The product must enable commanders to see and understand the operational environment in sufficient

---

[3] Cohen, Stephen, The Idea of Pakistan, Washington, Brookings Institution Press, 2004.

time and detail to employ their forces effectively. They must have the locations of friendly forces as well. Automation of the group's intelligence data base will provide near-real-time intelligence products to support both situation and target development.

Because their operational environment allows little margin for error, Special Forces teams must have detailed information about the operational area before they deploy. Special Forces mission success will directly depend on access to accurate, actionable intelligence in real time across the spectrum of intelligence agencies like NTRO, R&AW, NIA and IB. This should include real time satellite imagery and e-intelligence. For example, British SAS field commanders have direct access to MI 5 and MI 6 and Delta Force can pull down satellite images of latest and live enemy movements when actually in the process of executing missions. Additionally, the Special Forces commander can make an operation more effective if he determines its probable psychological impact in advance and then exploits it during the operation. If not analysed in advance, the particular operation may even produce a negative psychological impact on the population.

Peter Bergen in his book Manhunt[5] describing the 10 year search for Osama bin Laden and [6]Mark Owen giving the firsthand account of the Navy Seal mission that killed Osama bin Laden, have both described the intricacies, particularly intelligence gathering and its piecing together, of a modern day Special Forces mission. In the Indian context, the intelligence of the concerned area / location, contacts, communication, insertion, merging into surroundings, safe houses, execution of the mission, and extraction, all need meticulous intelligence, planning and execution. Conceptually, Special Forces should be employed to continuously 'shape the battlefield' from conventional wars in nuclear backdrop to asymmetric and fourth generation wars. Their employment should be theatre specific and including as force multipliers to complement tasks performed by conventional forces, entailing high risk, high gain missions having min visibility with desired effect. Their strategic tasking should be synchronized with national security objectives.

---

[4] US Army Field Manual 3-05.102, Army Special Forces Operations Intelligence, http://www.fas.org/irp/doddr/army/fm3-05-102.pdf

[5] Bergen, Peter, Manhunt, The Bodley Head, London, 2012.

[6] Owen, Mark with Maurer, Kevin, No Easy Day, Penguin Books Ltd, England, 2012.

Indian Special Forces would be required to undertake variety of tasks: information support operations; surveillance and target designation in areas of strategic interest; shaping asymmetric and conventional battlefield to India's advantage; deter opponents exploiting our fault lines; controlling fault lines of adversaries; undertake psychological operations, perception management and unconventional warfare; anti hijack; building partner capabilities with friendly countries; and providing cutting edge for strategic force projection.

The rank and file of Special Forces need not be rigid like the military. The British SAS has no formal rank structure and operates in civvies. Similarly, individual qualities required need not be uniform albeit basic qualities of being volunteer, physical and mentally toughness, initiative and innovativeness, understanding of technology and mission intent, abilities of split decision making, operating independently, blending with surroundings, optimising local resources, and proficiency in handling mission related arms, equipment, explosives will be assets.

The buzzword eventually has to be mission specific training. [7]Azam Gill, former Pakistani Army-ISI Officer, later Brigadier in French Foreign Legion and lecturer with French Navy, when interviewed by Geopolitics in March 2012 said this, "French Foreign Legion recruits bring different talents and experience to the Legion - beggars, doctors, army officers, racing car drivers, princes, bandits, locksmiths etc. when the Legion needs to put together a team for a special operation, all this prior talent come into play. ...... outperforms all French Army Units and is first to be launched in combat situation. The Legion has never claimed to be the best, says it is against the best, probably the best, recognises equals but knows no superiors. It does not acknowledge all operations." [8]His book, 'Blood Money' gives an insight into mercenary soldiering in the backdrop of a scenario strikingly close to that which may have led up to the attack on the Twin Towers.

Due to the hi-tech nature of the war against terrorism, there will be a need to attract highly skilled and technically qualified young civilians directly

---

[7] Gill, Azam, I Made Unauthorised Sorties Behind Indian Lines, *Geopolitics*, Vol II, Issue X, March 2012.

[8] Gill, Azam, Blood Money, BeWrite Books, Vancouver, Canada, 2002.

from IITs and computer institutes since the new war will be fought in the electronic and cyber dimension as well. Accordingly, there is a requirement to adapt and change recruitment policies with appropriate incentives so that fresh talent from outside the three services can be attracted. Young men who have special flair, creative and innovative minds or have unusual technical skills including in cyber warfare, e-warfare, cryptography, NBC, offshore drilling technology, specialists in the field of psychological operations, e-communications or language skills etc. Street-smart young men (and women), who can out-think others on their feet, conceptualize imaginatively and think out of the box. People who can do unexpected and unorthodox things with flair and élan should be encouraged to come on board for short duration or on deputation especially to form the non-uniformed or uniformed support structure.

Appropriately, specific Special Forces units/sub-units would need to develop specialist expertise and combat skills in specific core competencies, some of them being: counter-terrorism; counter-insurgency; unconventional warfare; counter-hijack; hostage rescue; cyber terrorism including e-warfare/propaganda, e-finance; electronic warfare; psychological operations and civil affairs; NBC warfare; special aviation quick reaction and counter-terrorism tasks specific to aviation environment; special marine and boat section quick reaction and counter terrorism tasks specific to a marine environment; special high altitude and snow warfare insurgency tasks; demolitions, explosives, industrial sabotage, offshore and land based oil installations and NBC disaster management; snatch operations and pseudo gang warfare.

Though threats have been discussed in an earlier chapter, it is prudent to review them here as they help shape what our doctrine and concept of employment for Special Forces should be. The emergent threat as it has unfolded during the last two decades is external and internal; latter mainly Maoist and sundry other insurgencies that are well known. It is characterised by a combination of low-level urban and rural insurgency of a non-conventional and asymmetrical nature, embedded with random but high profile acts of terrorism, hijackings, car-bombings, kidnappings and Maoist-style killings that are dispersed in time and space. The external threat is essentially from so called non-state actors mostly Pakistan sponsored including jihadi groups such as Al Qaeda, LeT, JeM et al, acting alone or in cahoots with Pak Army-ISI and Pak Taliban and backed by its regular forces. Add to this the threat from NBC proliferation and the gravity of the emerging threat can be readily assessed. The more recent Chinese involvement in supporting the Maoists has increased the gravity of the situation. As for the conventional, though the conventional threat from Pakistan across our Western front appears to be muted and can be dealt with, the newly emerging military nexus between Pakistan and China in the sensitive areas of The Northern Territories along the Baltistan – Gilgit – Skardu Axis and then onwards into Muzafarnagar – POK is alarming. This development has serious implications for India's defense posture not only for J&K but for entire North-Western India as well. To the East, the Chinese designs extend beyond the illegally occupied Aksai Chin to her designs on the Doklam Plateau in Bhutan and onwards to her claim of entire Arunachal Pradesh.

Counter measures need to be put in place on the diplomatic front by drawing attention of the UN, the US and other friendly Central Asian countries. With Siachen as a pivot, the area and the passes to its North would need to be brought under close observation and surveillance from space, air and ground. In this regard, long range Special Forces patrols appear to be the best option to keep a close watch on developments and intelligence gathering. Consequently, our Special Forces should initiate / intensify joint training to improve interoperability with countries of the Central Asian Republics, Afghanistan, Japan, South Korea, Vietnam, US and Australia. We must also work out a road map to increase our foot prints in Taiwan by formulating a Treaty of Peace, Friendship and Mutual Cooperation in the event of continued Chinese belligerence, territorial claims to more Indian Territory and support and arming of Indian insurgents in northeast and Maoists continues unabated. Taiwan is a reality and there is no reason for us not to expand relations with that country especially considering Chinese intransigence to the terrorism being perpetrated by Pakistan in India.

Unless we take into account the vulnerabilities of China and Pakistan in shaping our policy and in sowing the seeds of the architecture of our strategic co-operation with the US, Afghanistan, Central Asian Republics, Australia, Japan, South Korea, Vietnam and Taiwan, we may end up with a policy which is over-focused on traditional military aspects and under-focused on non-traditional aspects of internal frictions and fragilities in China[9]. China's increasing vulnerabilities should be a matter of core interest to us with respect to deterring her irregular war.

The nature of the new war is hi-tech with diverse technological dimensions. These cover the use of satellite phones and thermal imagery, cyber-warfare, cyber-terrorism, e- warfare, e-propaganda and e-finance (money laundering) to list just a few. For instance, it is well known that it was Pakistani born Khalid Sheikh Mohammad, who masterminded the attack on the Twin Towers on 9/11 and laundered money electronically from banks in Karachi to the hijackers in the US. He was also involved in the David Pearl assassination case. Interestingly, Khalid Sheikh was one of the four hardcore jehadi leaders released from Jammu prison in the aftermath of IA flight IC 814 fiasco in December 1999 together with Salha-uddin, Hafiz Saeed, and Omar Sheikh of Al Qaida who, on release, went on to plot other terrorist attacks on India starting from the Parliament attack. This dramatically highlights the type of linkages and synergy that exists between various jihadi groups worldwide. They were released because the party in power was risk averse to take bold decisions thereby unleashing a chain of terrorist attacks that are still rampant.

Glaringly, this also underscores the utter incompetence and lack of coordination of our intelligence and security establishment. It should, therefore, be self-evident that the sine qua non for countering the above type of asymmetrical, non- conventional threat, employing hi-tech methods, requires audacious risk taking and political decisions in real time. It requires joint planning, joint coordination and joint synergy of a very high order from the Services hierarchy and the political apex. Moreover, it also requires fresh thinking out of the box by exploring new approaches and by applying unorthodox means. Experience in US, UK and Israel has shown that this is possible by raising a customised 'lean and mean' type of Special Forces imbued with exceptional leadership qualities, specialised combat skills and robust aggressive spirit. A force that is small, light-footed and nimble enough

---

[9] Raman, B, Core Concerns, Core Interests, *outlookindia.com*, 03 March, 2012.

to execute quick surgical strikes yet apply deadly force with pinpoint precision. It is in this context that the restructuring of our Special Forces needs to be undertaken in focused manner, as recommended in the succeeding chapter.

5

Indian Special Forces – Circa 2030

What India needs for coping with irregular / unconventional threats and power projection is a set up as follows: *Special Forces* – for strategic tasks including as deterrent against irregular and asymmetric warfare; *Commando Forces* – for cross border tactical tasks and within border tasks beyond capabilities of regular infantry; and *Airborne Forces* – for rapid reaction and force projection within and outside India. Special Forces though primarily tasked for strategic tasks throughout the conflict spectrum will also act as force multipliers to Commando Forces and Airborne Forces during conventional conflict.

The Prime Minister appointed Naresh Chandra Committee has recommended a Permanent Chairman of the Chiefs of Staff committee (COSC). The Kargil Review Committee and the Group of Minister's (GoP) had recommended appointment of a Chief of Defence Staff (CDS). Notwithstanding the difference in nomenclature, a Permanent Chairman of COSC or CDS is going to be no different from his counterpart in Pakistan – sans requisite powers. This is more so in the Indian Context because if and when the CDS is appointed, he will have equal voting power on any issue as the Service Chiefs and in case of dissent by two Service Chiefs, the arbitration is to be done by the Ministry of Defence (MoD). CDS becoming 'Single Point Advisor' to government is therefore a misnomer.

Most significantly, strategic Special Forces missions in most countries are controlled and executed by the political authority without reference to even the highest military authority due to their politico-military nature and very high sensitivity. The issue of politico-military will, too requires to be addressed, which can only be done by the highest political authority. In the current dispensation, there have been times when covert missions have not taken off, though desired by the political authority, because of lack of institutionalised cover for the operatives. Details of such missions obviously cannot be mentioned here due to their sensitive nature. However, resolution of such issues can be found with requisite politico-military dialogue.

There has been much discussion and speculation on the structure that should come up for the proposed Special Forces Command in India. One suggestion is that this be created by integrating existing Special Forces under the CDS / Permanent Chairman of the COSC and resources from such command be drawn by the national hierarchy (through R&AW) for strategic tasks, when required. This thinking is based on Operation 'Neptune Spear' by Seal Team 6 in killing Osama-bin-Laden. Though Seal Team 6 is under US SOCOM, it was actually launched by the CIA. The effectiveness of such an arrangement is unworkable in India on two counts: the CDS / Permanent Chairman of the COSC, if and when appointed, will not have the same powers as Commander US SOCOM in deploying resources routinely including in multiple countries for continuous information support operations, strategic surveillance etc; and such arrangement will have to filter through various layers of the MoD with attenuated problems. It is prudent to note that while R&AW has been drawing upon the Special Groups of SFF for various tasks, we have not been able to create the required deterrence in any measure at all. Another school of thought is to put the Special Forces Command under the National Security Advisor (NSA) akin to the Strategic Forces Command (SFC). However, the Special Forces Command really is not within the ambit of the charter of the NSA either.

Keeping the above in mind, the Special Forces of India should be re-organised or raised, directly under the Prime Minister with the nucleus selected from Military Special Forces and an all India manpower base including Military, PMF and CAPF. They can be organised into and named "Indian Special Forces Command (ISFC)".

Commander ISFC should be on deputation or permanent absorption from Army Special Forces. Picking up a commander from a senior police officer akin to being done in case of NSG will be a serious mistake. Commander ISFC should have complete freedom in selection of manpower, weapons and equipping of his force. Manpower selection for Special Forces is a vital issue. It is significant to note how seriously the USSF takes the issue of quality manpower. Take for example personnel policies of SEAL Team 6, which carried out the[1] raid to kill Osama bin Laden. US Navy personnel volunteer for the SEALS and only those who make the cut are inducted. They serve in other SEAL teams for several years gaining operational experience. Out of this lot, volunteers then opt for SEAL Team 6, which means they are truly the best of the best. Their average age profile is 32 years, which shows they have the right mix of youth and experience. This also underlines the seriousness with which the US develops their Special Forces for various roles. The success of Op Neptune Spear is a testimony to that.

The organisation of ISFC to start with should be about two battalion worth that can be expanded upon subsequently. The initial groupings of Special Forces Teams (SFTs) may individually comprise of anything from 25 to 50 per country or region specific operatives, duly prioritized. The size of the individual SFT will depend upon the country / region and its relative importance in terms of national security objectives. They should have institutionalised access to integrated intelligence, varied insertion and extraction capability and adequate support elements. The Insertion-Extraction Group should have the means to deliver and extract the SFTs through air, surface and underwater. It would be prudent to commence specialised training and preparation with some 300-350 operatives only.

While direct action training can be common including advance specialist training at the Special Forces Training School (SFTS), additional training including language training will have to be specific to the area and manner of their employment for specific missions. A rigid rank structure should be avoided as should be the wearing set pattern of uniforms. It would be prudent for the ISFC to have a Cyber Cell to specifically monitor and target terrorist networks and propaganda. There should be an R&D element to customize state of art weapons and equipment to Special Forces requirements in general as well as for specific missions.

Special Forces should be strategically tasked for politico-military missions as follows: information support operations; surveillance and target

---

[1] Owen, Mark with Maurer, Kevin, No Easy Day, Penguin Books Ltd, England, 2012.

designation in areas of strategic interest.; shaping asymmetric and conventional battlefield to Indian advantage; deter opponents exploiting our fault lines; control fault lines of adversaries; undertake information / psychological operations and unconventional warfare; anti hijack; build partner capabilities with friendly countries; and provide cutting edge for strategic force projection.

The Prime Minister should have a Special Forces Cell comprising serving and veteran Special Forces, R&AW, NTRO and IB officers as an adjunct to the PMO; to act as the "Brain" to evolve a National Doctrine and Strategy for Employment of Special Forces, oversee their manning, equipping, training, consolidation, operational and intelligence inputs, inter-agency synergy and strategic tasking. This Cell should coordinate continuous all source intelligence gathering and automated analysis and assessments (short, medium and long term) supported by an automated decision support system and real time dissemination on required basis. The broad outline organisation of ISFC is recommended as under:

Outline Organization: Indian Special Forces Command (ISFC)

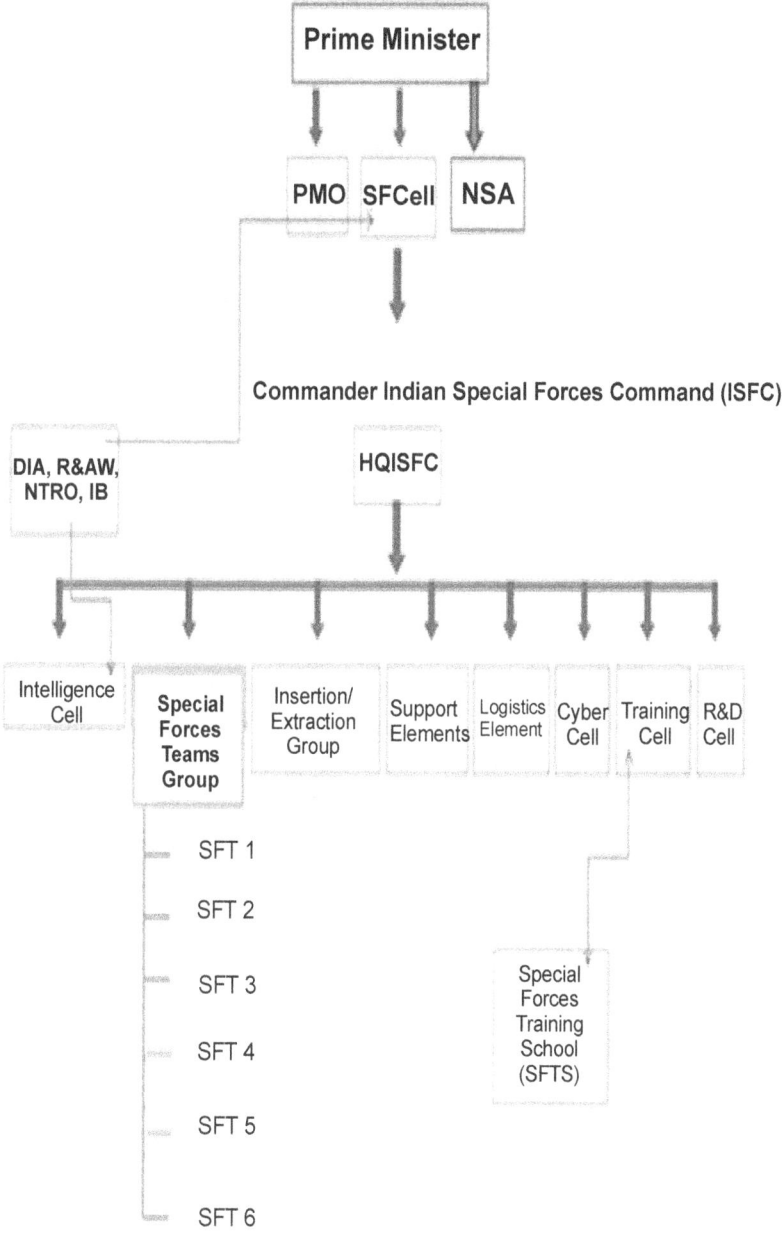

Prime Minister

PMO SFCell **NSA**

Commander Indian Special Forces Command (ISFC)

DIA, R&AW, NTRO, IB

HQISFC

Intelligence Cell

Special Forces Teams Group

Insertion/ Extraction Group

Support Elements

Logistics Element

Cyber Cell

Training Cell

R&D Cell

SFT 1

SFT 2

SFT 3

SFT 4

SFT 5

SFT 6

Special Forces Training School (SFTS)

Note: Number of SFTs and Composition of Individual SFT not sacrosanct.

The balance of what really are Commando Forces should be reorganised into an Integrated Commando Command (ICC) directly under the CDS / Permanent Chairman COSC - in line with what the Naresh Chandra Committee has recommended. The ICC should integrate the Army Special Forces, MARCOS, Garuds, NSG and SFF.

The Commander of ICC must be from Special Forces. The media has been talking of the Services raising three new Commands viz Cyber, Aerospace and Special Forces. There is a mention that the officer heading the Special Forces Command will be a three star rank officer. This will be problematic as a three star rank officer from Special Forces is a rare commodity in India though there have been exceptions where up to three Special Forces officers wore this rank at the same time. If this stipulation is followed rigidly, then the situation will be back to square one, as in the case of ADGMO (SF) in Military Operations Directorate – a non-Special Forces officer heading the ICC.

The ICC will axiomatically meet individual service requirements of the three services as well. Raising of a Marine Corps as part of the ICC and locating it in the Andaman & Nicobar Command (ANC) too needs to be given due consideration, which will also assist the ANC and dilute to some extent the present disadvantage of the ANC looking to the mainland for troops that may be unworkable in the emerging strategic environment given ANC's vast regional responsibility and possibilities of the IOR heating up. Navy's case for raising a Marine Brigade is lying with the government for over a decade now. As regards the SFF, with the Tibet born SFF personnel having retired long since, time is also opportune to actively consider regularizing the SFF. A Tibet origin girl in India has already won the court battle for being granted Indian nationality. The recommended outline organisation of the ICC is as under:

ine Organization of Intergrated Commando Command (ICC)

CDS/ Permanent Chairman COSC

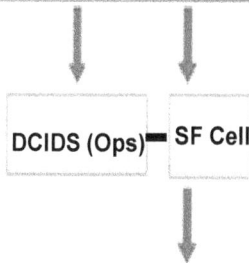

DCIDS (Ops)— SF Cell

ommander Integrated Commando Command (ICC)

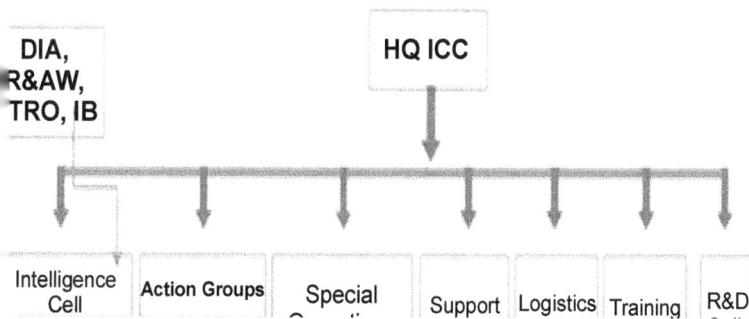

DIA, R&AW, TRO, IB		HQ ICC

| Intelligence Cell | Action Groups | Special | Support | Logistics | Training | R&D |

The very first time mention of a two front war came up in India was in a letter dated 7th November 1950 from Sardar Patel to Pandit Nehru,[2] extracts of which are, "Chinese irredentism and communist imperialism are different from the expansionism or imperialism of the western powers. The former has a cloak of ideology which makes it ten times more dangerous. In the guise of ideological expansion lie concealed racial, national or historical claims. The danger from the north and north-east, therefore, becomes both communist and imperialist. While our western and north-western threat to security is still as prominent as before, a new threat has developed from the north and north-east. Thus, for the first time, after centuries, India's defence has to concentrate itself on two fronts simultaneously. Our defence measures have so far been based on the calculations of superiority over Pakistan. In our calculations we shall now have to reckon with communist China in the north and in the north-east, a communist China which has definite ambitions and aims and which does not, in any way, seem friendly disposed towards us".

In the wake of China's aggressiveness in recent years, there has been talk of a two front war. More significantly, the Indian Army has been talking of a two and a half front war. China that was earlier using Pakistan as proxy for stoking internal fires in India has now, in addition, taken on a direct role in arming and equipping northeast terrorists groups and the Maoists. Post 2014, the situation in Afghanistan too will aggravate to India's disadvantage. [3]To reiterate what Agha H Amin, Pakistani defence analyst and former Pakistani army officer has said, "Utopians in India are jubilant that Pakistan has made peace with India. Nothing in reality can be farther from the truth. Pakistan's apparent shift is merely a tactical response to extreme confrontation with the US over perceived US view that Pakistan is playing a double game in Afghanistan. The real picture of true intentions of the Pakistani military will emerge when the US withdraws from Afghanistan. This will be the time when the Russians, Iranians and Indians will have no choice but to support the Northern Alliance against Pakistan sponsored

---

[2] Sardar Patel's Letter to Jawaharlal Nehru (7 November 1950): http://www.friendsoftibet.org/main/sardar.html

[3] Amin, H, Agha, Can India and Pakistan Make Peace,http://reportmysignalpm.blogspot.in/2012/08/india-and-pakistan-make-peace-by-major.html

Taliban who regard all Shias, Ismailis, non-Pashtuns, moderate Pashtuns as infidels who deserve to be massacred. Pakistani politicians will remain the puppets of the military that they have been since 1977. Terrorism will remain a tool of foreign policy while the Pakistani Military runs the Pakistani state under a facade of PPP or PML or Tehrik i Insaaf………..Pakistani military will be hoping to achieve all its objectives: an extremist dominated Afghanistan; a Baluchistan fully fragmented and crushed; a Pakistani political party leading Pakistan fully subservient to the Pakistani military; a renewed infiltration in Kashmir; a brinkman's nuclear policy with India; and a greater Chinese vassal with far greater Chinese interests in Pakistan. There is no doubt that Pakistan will be a semi autonomous Chinese province by 2030 or so. …. Pakistani Baluchistan by 2030 would be a completely Chinese run show… This means that Pakistan's political economy of exporting terrorism as a foreign policy tool, massive corruption at home and the resultant ever growing reservoir of economically deprived youngsters who will fill ranks of extremists and suicide bombers will continue".

Numerous indicators exist that the China-Pakistan nexus will up the ante of internal security situation in India and in case of conventional conflict, come up with surprises to convert this internal 'half front' to a full blown 'third front' akin to a situation of civil war. In addition, on the conventional war front too, we should expect asymmetric surprises particularly in the case of China. The discovery of a 1.5 Kilogram Uranium IED in Assam and Maoists expanding their area of operations to Assam since 2013 bodes the increasing dangers.Prudence demands that we holistically assess these threats and take adequate measures to nullify these asymmetries and more importantly tilt them to our advantage. Our Special Forces should be in the forefront of such measures.

Another important issue that needs to be urgently addressed is the civil-military understanding gap[4]; something that plagues most countries. A sample US view from Rosa Brooks says it all, "Most Americans know roughly as much about the U.S. military as they know about the surface of

---

[4] Brooks, Rosa, Generals Are from Mars, Their Bosses Are from Venice, Foreign Policy Magazine, 25 July, 2012.

the moon. It's not that we don't like the military — we love it! We just don't have a clue who's in it, what it does, what it costs us, or what it costs those who join it. And as a nation, we don't particularly care, either." Happenings over the years in India indicate continuing lack of required politico-military connect that has kept us at tremendous disadvantage strategically, impinging on national security. In fact, the panacea to the immense strategic deficit of India needs to be met not only by early appointment of the CDS but in addition, establishing Advisory Military Cells comprising of veteran and serving military officers directly under the Prime Minister, External Affairs Minister, Defence Minister, Home Minister and the National Security Advisor. The politico-military connect in any case is vital to setting up the ISFC and ICC as there will be heavy opposition from agencies that control the concerned forces presently viz the Cabinet Secretariat, Ministry of Home Affairs and the Services. However, the exercise is essential in the present and future context, akin to the efforts for integrating the nine major intelligence agencies and setting up the National Counter Terrorism Centre (NCTC). The fact is that diplomacy and conventional capability by themselves cannot contend with asymmetric wars of Pakistan and China.

Epilogue

On the night of May 1, 2012, two specially-fitted Black Hawk helicopters slipped into Pakistani airspace from across the border in Afghanistan and headed for a military cantonment town called Abbotabad. Named after a British General, the town was home to the prestigious Pakistan Military Academy where generations of post-independent army officers had been commissioned. The two Black Hawks had two Chinook Helicopters that would provide logistic support for a mission was so secretive that only a handful in the President Barack Obama administration knew about it. *Operation Neptune Spear* was ten years in the making, when several civilian aircrafts had been hijacked by the members of the *Al Qaeda* and rammed into the twin towers of the World Trade Center and the Pentagon, killing scores of innocent civilians.

The man who had planned that terrorist attack was Osama bin Laden, a Saudi Arabian national who had founded the terrorist organisation which, when translated into English meant 'The Base'. The *Al Qaeda* would become a symbol for terrorists across the world, as it began to wage a new war that few had seen and were untrained for. In fact, on the day the civil aircraft rammed into the twin towers on September 9, 2001, the U.S Military was preparing for a mock nuclear strike exercise. Few in the strategic community could imagine that a plane loaded with aviation fuel could be transformed into an instrument of mass destruction.

The answer to counter this elusive enemy was complex and offered myriad challenges. Unlike their past enemies, this new enemy did not hold ground, had little to lose, and therefore was difficult to deter and was a non-state actor. This made him elusive and gave him the strategic advantage of a successful first strike.

For the American military, adapting their Cold War doctrine to a new asymmetrical war became a primary challenge. But much of the new doctrines became possible because of an earlier special operation that had led to a spectacular failure. The failure of *Operation Eagle Claw* to rescue American hostages from Iran in the summer of 1980 had led to a major reorganisation and the birth of the Special Operations Command (SOCOM). The man who would plan *Operation Neptune Spear* was a much revered Special Forces officer, Admiral William McRaven, who had served with the Navy SEALS.

Admiral McRaven would send in SEAL TEAM 6 (also known as DEVGRU or Naval Warfare Development Group) a secretive unit that took the best of the men from the other SEAL units. The basic idea of SEAL TEAM 6 was to pick the best and the most experienced SEALS who would go through additional months of grueling training sessions before being accepted into this tiny and elite community. Admiral McRaven knew them well and was also recognised as a Captain earlier had authored the seminal study '*The Theory of Special Operations*' considered the Bible of Special Forces the world over. Admiral McRaven had argued that if Special Forces had to succeed in special operations they would have a small window of opportunity to gain (or lose) a point of superiority. This could be achieved if the operations were simple, daring and well planned and executed by highly-trained Special Forces with special skills. Their operations, he argued, would always be of a strategic nature for strategic gains and not smaller tactical victories. The Special Forces, he felt had to be used judiciously as game-changers in a war and had to be factored in as a major factor that would impact the overall course and the theatre of the war and not an isolated battle.

The US military also recognised that to fight this new war, they needed new doctrines and the old water-tight compartments would have to be broken down to forge new capabilities. Over the next few months the American intelligence community and the Special Forces community would recognise each other as natural allies, working closely together to take down this enemy. They would offer a strategic option to their political leadership that could achieve strategic gains that conventional forces were not designed or geared to gain.

In India, that faces multiple challenges, the current security architecture is moribund and static, lacking in the dynamism that nuclear-armed neighbours and cross-border terrorism has posed for decades. In the north-east, a rapidly modernising Chinese People's Liberation Army (PLA) recognised the need for separate rapid reaction forces and Special Forces and quickly developed both capabilities for every Military Region.

Unfortunately, the basic lack of understanding of the use of Special Forces in India has left them in a catch-22 situation. Used in roles that they are not suited for, gives the wrong impression to formation commanders who come from the conventional military. This under-appreciation leads to a vicious cycle that continues to exist to the detriment of our Special Forces. In 2005, while addressing a prestigious seminar in New Delhi, the then Union Defence Minister, Shri Pranab Mukherjee had pointed out that India's strategic requirements stretched from the Horn of Africa in the West to the Malacca Straits in the East with the Central Asian Republics and the China to its North. Unfortunately, barring an undefined strategic role for the Special Group that is part of the SFF, the other Special Forces units continue to stare at a limited role that is mostly tactical in scope. This is undesirable and detrimental for the strategic needs for a country of the size and potential that India has.

Special Forces will continue to dominate the whole spectrum of conflict for several decades and will be critical to shape its outcome. We can continue to ignore this axiom at our peril. Or, we can change things and begin a new chapter that will finally give the Indian Special Forces what they need to achieve what they are truly capable of. We hope that day is not too far away.

Bibliography

Books

1. Bank, Aaron, From OSS to Green Beret – The Birth of Special Forces, Presidio Publishers, USA, 1986.

2. Bergen, Peter, Manhunt, The Bodley Head, UK, 2012.

3. Qiao Liang, Qiao and Xiangsui, Wang, Unrestricted Warfare, PLA Literature and Arts Publishing House, Beijing, 1999.

4. Oberoi, Vijay, Special Forces: Doctrine, Tasking, Equipping and Employment, Centre for Land Warfare Studies, New Delhi, 2006.

5. Collins, John M, Green Berets, SEALs Spetznaz, Brassey's, International Defence Publishers, UK, 1989.

6. Praval, KC, India's Paratroopers, Vanity Books, Delhi, 1974.

8. Sabharwal, Maj Gen (Retd) OP, The Killer Instinct, Rupa and Co, 2000.

9. Karim, Maj Gen Afsir, The Story of Indian Airborne Troops, Lancer International, New Delhi, 1993.

10. Suvorov, Victor, Spetznaz: The Story Behind the Soviet SAS, Hamish Hamilton Ltd, 1987.

11. Simpkin, Richard E, Race to The Swift, Brassey's Defence Publishers Ltd, UK, 1985.

12. Col Collins, John M, Special Operations Forces: An Assessment, National Defense University Press, USA, 1999.

13. Musa, General Mohammed, My Version, Oxford University Press, Pakistan 1993.

14. Beckwith, Charles, Knox, Donald, Delta Force, Arms and Armour Press, UK,   1984.

15. Marwah, Ved, Uncivil Wars: Pathology of Terrorism in India, Penguin India,   1997.

16. Farran, Roy, Winged Dagger, Arms and Armour Press, UK, 1984.

17. Salik, Siddiq, Witness to Surrender, Oxford University Press, 1882.

18. Lang, Walter N, The Worlds Elite Forces, Salamander Books Ltd, UK, 1987.

19. Adam, James, Secret Armies: The Full Story of SAS, Delta Force & Spetznaz, Century Hutchinson Ltd, UK, 1987.

20. Crawford, Steve, SAS Gulf Warriors, Simon & Schuster Ltd, 1996.

21. Ransom, Harry, Rowe, Central Intelligence and National Security, Harvard University Press, USA, 1958.

22. Paschal Rod, LIC 2010 Special Operations and Unconventional Warfare in Next Century, Brassey's, Inc, USA, 1990.

23. McNab, Andy, Bravo Two Zero, Bantam Press, UK, 1998.

24. Connor, Ken, Ghost Force: The Secret History of the SAS, Cassel & Co, UK, 2002.

25. Beaumont, Roger A, Military Elites, Robert Hale & Co, London, 1976.

26. David, Alexander, Special Operations, Berkley Publishing Group, Penguin Putnam inc, USA, 2001.

27. Kermit, Roosevelt, War Reports of the OSS, Walker & Co, USA, 1976.

28. Gavin, John R, General, Air Assault: The Development of Airmobile Warfare, Hawthorne Books, USA, 1969.

29. Arostegui, Martin C, Twilight Warriors: Inside the World's Special Forces, St Martin's Press, USA, 1995.

30. Conboy, Ken, Elite Forces of India and Pakistan, Osprey Press, UK, 1992.

31. Kelly, Ross, Special Operations and National Purpose, Lexington Books, USA, 1989.

32. Plaster John, SOG: The Secret Wars of America's Commandos in Vietnam, Simon & Schuster, USA, 1994.

33. Aijazuddin, FS, From a Head, Through a Head, To a Head, Oxford University Press, Karachi, Pakistan, 2000.

34. Dhar, Krishna, Maloy, Open Secrets – India's Intelligence Unveiled, Manas Publications, New Delhi, India 2005.

35. Jaggia, Anil K and Shukla, Saurabh, IC-814 Hijacked: The Inside Story, Roli Books, New Delhi, 2000, pp 59-60.

36. Sen, Lionel, Pratip, Slender Was the Thread : Kashmir Confrontation 1947-1948, Orient Longman Ltd, New Delhi, 1969.

37. Owen, Mark with Maurer, Kevin, No Easy Day, Penguin Books Ltd, England, 2012.

38. Cohen, Stephen, The Idea of Pakistan, Washington, Brookings Institution Press, 2004.

39. Gill, Azam, Blood Money, BeWrite Books, Vancouver, Canada, 2002.

Articles / Web Reference

1. http://www.fas.org/sgp/crs/natsec/RS21408.pdf

2. http://www.soc.mil/USASOC%20Headquarters/SOF%20Truths.html

3. National Security Guards, http://www.bharat-rakshak.com/LAND-FORCES/ Special-Forces/NSG.html

4. Kasturi, Bhashyam, National Security Guards: Organisation, Operations and Future Orientations, Indian Defence Review, Vol. 8 (3), October 1993, pp. 59 -63.

5. Kasturi, Bhashyam, National Security Guards: Past, Present and Future, Bharat Rakshak Monitor, Volume 5 (5), March, 2003. http://www.bharat-rakshak.com/MONITOR/ISSUES-5/Kasturi.html

6. Special Frontier Force, Wikipedia, http://en.wikipedia.org/Special-Frontier Force

7. Photos Special Frontier Force, http://www.militaryphotos.net/forums/showthread.php?62282-very-rare-pics-of-some-of-india-s-Special-Forces-speciality-units

8. India: The Next Super Power, The Economic Times, March 7, 2012 http://articleseconomictimes.indiatimes.com/2012-03-07/news/31132262_1_ superpower-state-hillary-clinton-Ise

9. Overdorf, James and Teng, Poh, India: Illegal guns plague cities, Globspot December 20, 2010. http://www.globspot.com/dispatch/India/101214/India-   illegal-guns-gun-control-crime

10. World Drug Report 2012, http://www.undoc.org/undoc/en/data-and-analysis/WDR-2012.html

11. India Ranks 134 in human development index, Hindustan Times, September 30, 2012. http://www.hindustantimes.com/News-Feed/India-ranks-134-in-human- development-index/Article1-76401.aspx

12. India 95th among 183 countries in Corruption Perception Index, The Economic Times. December 1, 2011. http://articles.economictimes.indiatimes//2011-12-01/news/30412959_1_corruption-perception-index-ranks-countries-cpi

13. Yusuf, Huma, Quest for a Dengue Free Democracy, The Dawn, October 7, 2011.

14. Hoodbhoy, Pervez, An Extremist Takeover of Pakistan is Probably NoFurther than Five to 10 Years Away, Dawn, February 26, 2011. http:// www.3quarksdaily.com/3quarksdaily/2011/02/an-extremist-takeover-ofpakistan-is-probably-no-further-than-five-to-10-years-away.html

15. Hoodbhoy, Pervez, The Suadi-isation of Pakistan, http://groups.yahoo.com/ group/Writers_Forum/message/39644

16. Beyond Bullets: Strategies for Countering Violent Terrorism, Study by the Centre for A New American Security http://www.cnas.org/files/

documents/ publications/LordNag Rosen- Beyond %20Bullets
%20edited%20Volume- june09_0.pdf

17. Tellis, J, Ashley, The Menace That Is Lashkar-e-Taiba, Policy Outlook,
    March 2012, http://carnegieendowment.org/2012/03/13/menace-that-
    is-lashkar-e-taiba

18. Riedel, Bruce, Mumbai terror Attack Group Lashkar-e-Tayyiba Now
    More dangerous Than Al Qaeda,http://www.dailybeast.com/articles/
    2012/07/02/ mumbai-terror-attack-group-lashkar-e-taiyyaba-noe-more-
    dangerous-than-al-qaeda. html/

19. Shah, Ali, Murtaza, Secret Pakistan, Dawn, October 27, 2011.

20. Amin, H, Agha, Can India and Pakistan Make Peace, http://
    reportmysignalpm.blogspot.in/2012/08/india-and-pakistan-make-peace-
    by-major.html

21. Watson, Joseph, Paul, US Special Forces Mass on Syrian Border,
    infowars.com, 13 December, 2011, http://www.infowars.com/us-
    special-forces- mass-on-syrian-border/

22. Watson, Joseph, Paul, Whether America Shares Its Values with
    Terrorists, prisonplanet.com, http://www.prisonplanet.com/

23. Joint Doctrine for Special Forces Unveiled, The Hindu, 02 October,
    2008, http://www.hindu.com/2006/10/2/stories/2008100256071 30. htm

24. US Army Field Manual 31 – 20, Doctrine for Special Forces Operations,
    http://www.enlisted.info/field-mannuals/fm-31-20-doctrine-for-special-
    forces-operations.shtml

25. US Army Field Manual 3-05.102, Army Special Forces Operations
    Intelligence, http://www.fas.org/irp/doddr/army/fm3-05-102.pdf

26. Gill, Azam, I Made Unauthorised Sorties Behind Indian Lines,
    Geopolitics, Vol II, Issue X, March 2012.

27. Sardar Patel's Letter to Jawaharlal Nehru (7 November 1950): http://
    www.friendsoftibet.org/main/sardar.html3.

28. Amin, H, Agha, Can India and Pakistan Make Peace,http://reportmy

signalpm.blogspot .in/2012/08/india-and-pakistan-make-peace-by-major.html

29. Brooks, Rosa, Generals Are from Mars, Their Bosses Are from Venice, Foreign Policy Magazine, 25 July, 2012.

30. Raman, B, Core Concerns, Core Interests, outlookindia.com, 03 March, 2012.

Index

W

Z

www.ingramcontent.com/pod-product-compliance
Lightning Source LLC
Chambersburg PA
CBHW060839100426
42814CB00016B/424/J